C.S. LEWIS' WAR
AGAINST SCIENTISM AND NATURALISM

JERRY BERGMAN

It is difficult to imagine a more popular or influential Christian apologist of the last century than C. S. Lewis, certainly among English speaking evangelicals. Through books such as *Mere Christianity, The Problem of Pain, The Great Divorce*, and numerous others, Lewis helped to define and re-energize the Christian message in the post-modern world. His allegorical works of fiction, from *The Chronicles of Narnia* to his Space Trilogy, remain enormously popular even today.

Because of Lewis' widespread popularity and influence, it has become essential that his position on the creation/evolution debate be clearly defined and established. Proponents of various movements, including theistic evolutionists and other "old earth" adherents, have been "jostling for a position at the lectern", so to speak, for Lewis' blessing on their particular position. For this reason, the importance of this book by Dr. Bergman cannot be overstated. The sole focus of this book is on Lewis' conclusions regarding this very crucial origins issue.

Through exhaustive research and documentation, Dr. Bergman has done an outstanding job of clarifying from Lewis' own writings and correspondence, and the reactions of his colleagues, the fact that he was without question an anti-Darwinist. Dr. Bergman accomplishes this first by defining Lewis' terminology in its proper context. (Certain terms and phrases from some of Lewis' earlier works have been incorrectly used by others out of context in an attempt to say that Lewis was an evolutionist.) Secondly, Dr. Bergman in a well-reasoned approach shows why Lewis avoided publicly criticizing evolutionary biology, all the while becoming more and more aware and convinced of the theory's philosophical senselessness and danger.

Finally, Dr. Bergman in this review of Lewis unveils previously unavailable documents (some posthumous) that reveal Lewis' complete rejection of what Lewis called the evolutionary "myth".

This is a must read for anyone, like myself, who has been inspired and enlightened by the works of C. S. Lewis and would therefore have great interest in Lewis' position regarding the critical issue of origins.

Kirk Toth Biologist, editor

Amazon Reviews of *C. S. Lewis: Anti-Darwinist: A Careful Examination of the Development of His Views on Darwinism*

If you're wondering what C. S. Lewis thought about Darwin and evolution, you have a choice: work through Lewis' voluminous output yourself, or read this short (129-page) book. Dr. Bergman has gathered the most important quotes by Lewis into 19 short chapters by subject. Was Lewis a theistic evolutionist? (No.) Was he a Young-Earth Creationist (YEC)? (unknown) Did he believe in the Fall of man and Adam and Eve? (Yes.) These and other topics are explored and supported with many quotations from Lewis and others. Noted Lewis scholar Louis A. Markos, who taught a course on C. S. Lewis for The Teaching Company, gives this book high marks, saying, "This is absolutely first rate; the best thing I've read on the subject."

Bergman explains why many misunderstand Lewis' views from quotes taken out of context. One needs to realize, as Bergman documents, that Lewis changed his views over time, becoming more staunchly anti-Darwinian in his later years. Because Lewis was not a scientist, and fighting Darwinists was not his primary mission in apologetics, Lewis used synonyms or indirect references rather than the word "evolution" for certain occasions. His most powerful objections were against naturalism and scientism. A good companion book on that subject is *The Magician's Twin* (ed. John West, Discovery Institute, 2012), which goes into Lewis' argument from reason in more detail. In short, another gem from Dr. Jerry Bergman, and I really like the Narnian twist on the "March of Progress" icon on the cover.

Dave C.

Jerry Bergman has done a wonderful service in looking deeply into the faith and writings of C. S. Lewis. Because of his popularity with a large group of people for many years, he is claimed as a stalwart supporter of old-Earth interpretations of Biblical chronology. Dr. Bergman is careful to show the humble unwillingness of this scholar to make pronouncements in scientific areas in which he wasn't an expert, but at the same time to show his disdain for Darwinism and support for truth as revealed in Scripture. Although he lived too late to benefit from the tremendous advances made by creation scientists in the recent past, he was yet unwilling to submit his thinking to the pressures of his secular peers. He felt the sting of persecution by Oxford dons for his Christian stand and the popularity of his writings, but was protected from censure by a move to Cambridge, orchestrated by a friend. It's amazing how such claimed bastions of free thought and investigation are often centers of secular intolerance for all things related to adherence to the things of Christ as revealed in the Bible; intolerance hidden in claims of the opposite.

Derek E. Iverson

Excellent book that pulls from many resources confirming C. S. Lewis' brilliant anti-Darwinist mechanistic evolution stance, and arguments and support for creation and intelligent design perspectives. I'm sure if Lewis were alive today with all the development of creation science, Biblical archaeology, and intelligent-design research, that Lewis would be all-the-more steadfast in his conviction of the Biblical historical truth found in Genesis.

Rev. A. Craig Bracy

As a college student in 1970, I was invited by a group of fellow students to spend an afternoon at Bethel Manor, a Christian co-op near the Michigan State University campus. Each one of us chose a book to read all in one sitting and then to discuss it in the evening. The book I chose was *Mere Christianity* by C. S. Lewis. That book, along with many others of his that I have read since, helped to establish a solid Christian foundation for my life. I am now on the board of directors for Bethel Manor, now called Living Rock Christian Co-op. Lewis, with his engaging writing style and solid logic in his understanding of the Christian faith and theology, is one of the most popular and influential Christian apologists of his day. Dr. Jerry Bergman is such a writer with similar stature in his thorough research of the subject, unearthing within the writings of Lewis the solid belief that Darwinism comes up short in its explanation for the origin of the world. Lewis kept his anti-Darwin ideas close to his chest, only to be revealed in the characters portrayed in his novels. Behind the scenes, though, were battles with those who opposed his Christian philosophy and resulted in exchanges with some of the more famous evolutionists of the day. Many of those writings were published posthumously, such as "A Reply to Professor Haldane" and "The Funeral of the Great Myth."

Dr. Bergman thoroughly refutes the idea that Lewis was a theistic evolutionist. Lewis expressed doubts and concerns about Darwinism throughout his writings.

Douglas B. Sharp

Good book! Dr. Bergman relates evidence from C. S. Lewis' writings to show that he rejected Darwin's theory of evolution. While doing so, Dr. Bergman provides a ton of support for us to consider in our own beliefs. The book is filled with descriptions of the reasons C. S. Lewis rejects macroevolution and documents the development of his beliefs from early dismissal of its importance to coming to accept it as central in a "web of falsehoods." He specifically discusses the difficulty in dealing with feeling unqualified to argue details with scientists and explains how to counter this fear. He talks about the danger of hypotheses being accepted as fact by the "climate of opinion," leading to tyranny against those who have opposing beliefs. He also talks about the potentially dangerous conditions reached by applying evolution theory in government policies. This is a relatively short book but it is filled with potent facts and thoughts. It is a highly recommended read.

TJ

When I read Lewis' works, I find myself "slogging" through them because of his big vocabulary and big IQ. Also, I am used to seeing creation vs. evolution from a "scientific" perspective and, since Lewis was not a formally educated scientist, he often presents a new approach for me and puts me in unfamiliar territory. Therefore, I am very grateful to Dr. Bergman for going through his works, ferreting out the pertinent information, "editing" it, and organizing it in a logical way so that I can begin to grasp what he has to say and where he was going (the only real evolution here seems to be his maturing stance against macroevolution). This must have been a monstrous task. At the end I asked myself why I felt what Lewis believed was so important. Without overthinking my answer, I feel that Lewis is one of the heroes of Christianity. He lived in the lair of higher education, and did not succumb to its propaganda but, by continuing to swim against a strong current against him, arrived at real truth (as is described in the last paragraph on page 84 of the book). At this point, Ben Stein's movie "No Intelligence Allowed" comes to my mind. I also think of the movie about Lewis, "Shadowlands," which presents him in a very favorable light.

Bryan Stuart

Some believe C. S. Lewis believed in evolution. However, the situation is more complicated as Dr. Bergman shows in this highly-documented book. Lewis had questions about evolution from even before he became a Christian, but because he was not a scientist, he was reluctant to address the subject in print. So, he made some statements early in his writing career that could be construed as he was favorable to Darwinism, especially if those statements are taken out of context. Later in his life he became more antagonistic toward evolution and even wrote a few essays about it. He was especially concerned about the moral ramifications of evolution. This attitude especially shows up in his private letters, which Bergman has analyzed. It is overall a good read and a strong case that Lewis was anti-Darwinian.

Mike Oard, M.S.

Endorsements

Dr. Bergman, in his inimitable style, has surfaced otherwise suppressed or not-easily-attainable historical information regarding C. S. Lewis' personal stance on creation vs. evolution. It is of tremendous importance where each of us, individually, "hangs our hat" in regard to these foundational questions and worldviews in life. C. S. Lewis opened many people's eyes in his generation and now beyond, to the tantamount urgency of taking the time to ponder and think on things of eternal significance. This new book by Dr. Bergman will help you understand how the great writer and theologian, C. S. Lewis, eventually came to his terms with "hanging his hat" on the truth that we are all accountable to a Divine Creator.

Bryce Gaudian, Hayward, Minnesota

What a happy surprise this book is! I was delighted to learn that C. S. Lewis rejected Darwinism in all its forms – certainly not a given in his day – before the modern creation-science movement had begun amassing evidence that the Genesis account of our origins is true. Dr. Bergman has done a masterful (and highly entertaining) job of documenting Lewis' unfolding thoughts on the subject, and proving that he was indeed an increasingly belligerent opponent of evolution theory.

Kitty Foth-Regner, author of *Heaven Without Her:*
A Desperate Daughters Search for the Heart of Her Mother's Faith

For many years, I assumed C. S. Lewis had made an uneasy truce with modern evolutionism. My opinion was based on certain passages in *Mere Christianity* and *The Problem of Pain*, in which he drew on evolutionist theories to illustrate Christian beliefs. It always seemed that one as committed to the Gospel as C. S. Lewis could not have been neutral on the issue of creation *ex nihilo* versus evolutionism. Lewis debunked modern arrogant unbelief based on evolution and guess-work which masquerades as "science" in his largely autobiographical *The Pilgrim's Regress*, the first book he published after his conversion. He also made it clear that what he had in mind in what he called emergent evolutionism, or developmentalism, is today called Darwinism. He put the following words in the mouth of foolish old "Mr. Enlightenment": "if you make the same guess often enough it ceases to be a guess and becomes a Scientific Fact." How appropriate these sarcastic words are now, fifty years later, about evolutionist dogma.

Lewis' essay, "Two Lectures," can only be called creationist apologetics. It contrasts an urbane lecture about "Evolution, development, the slow struggle upwards and onwards from crude and inchoate beginnings towards ever-increasing perfection and elaboration that appears to be the very formula of the whole Universe" with a Dream Lecturer pointing out that, as the "acorn comes from a full-grown oak, the Rocket comes, not from a still crude engine, but from something much more perfect than itself and much more complex, the mind of a man, and a man of genius." Bergman's book has proved beyond question that my early conclusions about Lewis were correct.

Ellen Myers, M.A.: Author and Teacher

C.S. LEWIS' WAR

AGAINST SCIENTISM AND NATURALISM

A DETAILED EXAMINATION
OF HIS VIEWS

Lead us, Evolution, lead us
Up the future's endless stair:
Chop us, change us, prod us, weed us,
For stagnation is despair:
Groping, guessing, yet progressing,
Lead us nobody knows where.

JERRY BERGMAN

Dedicated to David Bassett
Teacher, Scholar and a Dedicated Human Being

www.cantaroinstitute.org

C. S. Lewis' War Against Scientism and Naturalism:
A Detailed Examination of His Views
by Jerry Bergman

Published by Cántaro Publications, a publishing imprint of the Cántaro Institute, Jordan Station, ON.

© 2023 by Cántaro Institute. All rights reserved. Except for brief quotations in critical publications or reviews, no part of this book may be reproduced in any manner without prior written consent from the publishers.

Book design by Cántaro Institute

Library & Archives Canada

ISBN 978-1-990771-30-9

Printed in the United States of America

Table of Contents

Preface *by Karl Priest, M.A.* I

1 Introduction 1
    What Could One More Book Add 3
    Defining Terms 4
    The Reference Problem 8

2 Lewis' Terminology for Darwinism 13
    Myths about the Word Myth 16

3 Why C. S. Lewis Is Important Today 21
    Lewis and Education 26

4 What Made Lewis 31
    His Brother Warren 33
    Marriage 35
    His Death 36

5 Lewis' Educational Background 39
    His Profession as a Professor 43
    Lewis Faces Religious Discrimination 43
    Lewis, A Respected Oxford Scholar 46
    His Mission to Bring Christianity to the
    post-Christian World 46
    His Concern about Evolution Surfaces 49

6 Lewis Enters and Leaves Atheism	53
Lewis' Opposition to Eugenics	58
7 Return to Christianity and Opposition to Darwinism	65
8 Enter Owen Barfield	75
Lewis' Aggressive Opposition Against Darwinism	77
Other Evidence of Lewis' Anti-Darwinism	82
9 G. K. Chesterton's *The Everlasting Man*	89
What the Book Is About	92
10 Lewis Openly Rejects Naturalism	97
Why Lewis Opposed Naturalism	99
11 The Influence of Arthur J. Balfour	103
Balfour's Ideas Reflected in Lewis' Novels	108
Conclusions	112
12 The Darwinism that Lewis Rejected	115
13 Lewis' Answer to Darwinism	123
Lewis' View of Science	128
Common Interpretations of Lewis' Views of Evolution	129
14 A Harvard Scholar's View of Lewis	135
15 The Wielenberg Argument Against Darwinism from Good Design	143
16 Lewis Rejects the Theistic Evolutionism of Henri Bergson	153

17 Lewis Opposes the Theistic Evolutionism of Pierre Teilhard de Chardin	163
Summary	174
18 Lewis Opposes the Foundation of Evolution	179
19 Lewis Anticipated Arguments Used by Modern Evolutionists	185
Correspondence with Arthur C. Clarke	186
20 Lewis Teaches a Creation Worldview	193
Evolution as Mental or Spiritual Growth	195
21 Statements Indicating Lewis Was a Theistic Evolutionist	201
The Penelope Letter	207
Interpreting the Letter	208
22 Lewis and the Genesis Fall	213
Man Physically Descended from Animals	215
Analysis of the "Not Unlikely Tale"	224
23 Evolution Cannot Explain the Origin of the Mind or Life	231
The Helen Gardner Event	232
Lewis Discusses the Fall in *The Problem of Pain*	234
The Fall and Life in Outer Space	237
24 Understanding Lewis' View of Evolution – Continued	243
Lewis' Clear Statements Against Darwinism	245
Lewis' Growing Problem with Evolution	248
25 Lewis Becomes a Militant Anti-Darwinist	253

	Darwinism Versus Variation Within the Genesis Kinds	258
26	Lewis' Increasingly Strident Opposition to Evolutionary Naturalism	267
27	The Funeral of a Great Myth	275
	Lewis Not Anti-Science but Anti-Scientism	277
	Twisting the Facts	284
	The Watson Quote	287
	Darwinian Evolution Has Never Been Observed to Occur	293
28	The Canfield Letter	299
29	Lewis Supports the Implications of a Creation Worldview	307
	Atheist Recognizes Lewis' Rejection of Darwinism	309
	The Commercial World Welcomes the Myth	311
30	Lewis Censors His Anti-Evolution Views	317
31	Lewis' Concern Fulfilled in the Movement Against Anti-Evolutionists	331
32	Letter on the Bible Book of Genesis from Janet Wise	339
33	Some Conclusions	349
	The Anti-Darwinian Argument During Lewis' Lifetime	357
References		359

Appendix 1: A View from Another C. S. Lewis Fan 375

About the Author 381

Preface

by Karl Priest, M.A.

As I read his latest work, *C.S. Lewis: Anti-Darwinist*, I thought that Dr. Bergman is more than just a "mild-mannered" professor. He has carefully documented that Lewis had major concerns both about Darwinism and the potential harm that science could cause in modern society. Lewis has succeeded as have few others in causing Christianity to be discussed both seriously and publicly.

Lewis is the number one best-selling Christian author in the last century and has contributed greatly to Christianity being discussed academically. So far, four full-length movies have been made of his life[1] and many of his novels have been made into very successful, award-winning motion pictures. Lewis used Intelligent Design as a basis for his apologetics and, like any Christian who does not hide his light under a bushel, Lewis was the target of God-haters. Dr. Bergman robustly rebuts those who claim that Lewis believed in evolutionism. With powerful documentation, Dr. Bergman shows that C.S. was not even remotely a 'True Believer in Evolutionism.'

Some confusion as to where Lewis stood was created by his use of the word "evolution" in the context of spiritual or mental advance-

---

1. These include *The Life of C.S. Lewis* (Day of Discovery); *C.S. Lewis: Beyond Narnia* (Faith & Values Media) and the most successful *Shadowlands*, starring Anthony Hopkins and Debra Winger (Price Entertainment).

ment. Examination of some of his statements in context, and accounting where Lewis was in his spiritual growth, exclude the possibility that he ended up believing in theistic evolutionism. The more Lewis learned, the greater was his hostility toward evolutionism.

Nonetheless, "it is clear, however, from letters and essays that remained unpublished during his lifetime, that Lewis did much reading and thinking in private about evolution."[2] Perry notes, until the 1950s he tried to find a middle ground, accepting "the biological theorem' while rejecting its 'metaphysical statements.'" Actually, as I will document, he had problems with the theory even before he became a Christian. Perry continues writing, "in the 1950s he grew more skeptical [of evolution]. Between 1944 and 1960, he corresponded privately with Bernard Acworth (1885-1963), one of Britain's leading anti-evolutionists. In a 1951 letter, he noted that his doubts about evolution were being stimulated by the 'fanatical and twisted attitudes of its defenders.'"[3]

Lewis was reared as a nominal Christian, but after his mother's death professed atheism when he was only fifteen years old. His atheism was based on what he believed were the findings of science, specifically those that seemingly supported Darwinism. Not being a scientist, he had to take those findings on faith, actually on the authority of the scientists. He was also very influenced by a certain "freethinker" (a less confrontative term than atheist) teacher.[4]

He was at this time living, like so many people today, in a whirl of contradictions. He not only maintained that God did not exist, but was also very angry with God for not existing. He was equally angry with God for creating our world. Although he had doubts about evo-

---

2. Perry, Mike 1998 in Schultz and West's (1995) *The C. S. Lewis Reader's Encyclopedia*. Grand Rapids, MI: Zondervan, p. 69.
3. Perry, 1998, p. 69.
4. Joshi, S.T. 2003. *God's Defenders*. Amherst, NY: Prometheus Books, p. 108.

lution all along, after Lewis became a Christian he began to seriously question evolution.

Lewis was restrained in his public attacks on evolution because he did not have formal scientific credentials or training, and felt his attempts could hamper his main ministry of Christian apologetics. Like those who reject Darwinism today (especially scientists), Lewis worked in a hostile environment and came to take the bitterness and rancor due to his opposition as a matter of course. Lewis correctly knew that modern science was based on God as the Creator, writing: "Men became scientific because they expected Law in Nature, and they expected Law in Nature because they believed in a [law] Legislator" namely a Creator God.[5] Lewis added, "in most modern scientists this belief has died: It will be interesting to see how long their confidence in uniformity survives it."

Lewis was not anti-science; rather, he was anti-scientism. Scientism is the belief that modern science supplies the "only reliable method of knowledge about the world, and its corollary that scientists have the right to dictate a society's morals, religious beliefs, and even government policies merely because of their scientific expertise."[6] Lewis' concern was that the "words 'science' and 'scientific,' which had been given an inflated importance in the Victorian period, [had] swollen yet further to embrace all truth and knowledge. Questions of aesthetics, morality and, above all, questions of religion were relegated to the scrap heap where language was meaningless."[7]

For this rationale, using his skills as a writer and thinker, Lewis ag-

---

5. Lewis, C.S. 1947. *Miracles: A Preliminary Study*. London: HarperCollins, p. 110.
6. West, John G. (editor). 2012. *The Magician's Twin: C. S. Lewis on Science, Scientism, and Society*. Seattle, WA: Discovery Institute Press.
7. Wilson, A. N. 1990. *C.S. Lewis: A Biography*. New York: Norton, p. 86.

gressively and consistently attacked scientism and evolutionism. Lewis' most explicit anti-evolutionist writing is found in his essay "The Funeral of a Great Myth," the bulk of which deals with the mythological implications of evolutionism in a manner that is devastating to believers of such myths. Lewis makes it very clear that evolution is merely a hypothesis, and he speculates about the grounds for accepting the hypothesis as largely metaphysical and the fulfillment of an "imaginative need." Lewis added sarcastically that "probably every age gets, within certain limits, the science it desires." He used the term "emergent evolution" to describe Darwinism because the theory purported to explain the emergence of new life-forms. He called it "pure hallucination" in his essay "The Funeral of a Great Myth."

> If the solar system was brought about by an accidental collision, then the appearance of organic life on this planet was also an accident, and the whole evolution of Man was an accident too. If so, then all our present thoughts are mere accidents—the accidental by-product of the movement of atoms. And this holds for the thoughts of the materialists and astronomers as well as for anyone else's. But if *their* thoughts—i.e., of materialism and astronomy—are merely accidental by-products, why should we believe them to be true? I see no reason for believing that one accident should be able to give me a correct account of all the other accidents. It's like expecting that the accidental shape taken by the splash when you upset a milk jug should give you a correct account of how the jug was made and why it was upset.'[8]

In other words, Lewis concluded that if evolution were true, once the Solar System existed, then the appearance of organic life on this planet was ultimately an accident brought about by accidental collisions, and the whole evolution of Man was an accident, too. If so, then all our present thoughts are ultimately the accidental by-product of the

---

8. Lewis, C.S. 1984. *The Business of Heaven: Daily Readings from C. S. Lewis*. London, UK: Fount Paperbacks, p. 97. (Edited by Walter Hooper.)

movement of atoms. Thus, he concluded, "I see no reason for believing that one accident should be able to give me a correct account of all the other accidents."[9]

Lewis called evolution "the central and radical lie in the whole web of falsehood that now governs" modern civilization.[10] True Believers in Evolutionism believe that evolution explains everything, the stars, the galaxies, the Solar System, the planets and all life from amoebae to humans. Lewis explained that evolution is "not the logical result of what is vaguely called 'modern science,' but rather is a picture of reality that has resulted, not from empirical evidence, but from imagination (and) most people believe in orthodox evolution on the basis of authority 'because the scientists say so' and not on the basis of fact and scientific knowledge."[11]

Using common sense logic, Lewis pointed out that the adult human being was once an embryo, but the life of the embryo came from two adult human beings called parents. He added, "Since the egg-bird-egg sequence leads us to no plausible beginning, is it not reasonable to look for the real origin somewhere outside (of the) sequence altogether?"[12] You must go outside the sequence of inventions into the world of men, to find the real originator of the invention, asking, "Is it not equally reasonable to look outside Nature for the real Originator of the natural order?"[13] He realized that there was never a time when nothing

---

9. Ibid.
10. Lewis, C.S. Letter to Bernard Acworth (September 13, 1951) in *The Collected Letters of C. S. Lewis*, 2007, Vol. 3, p. 138.
11. Lewis, C.S. 1967. *Christian Reflections*. Grand Rapids, MI: Eerdmans, p. 103. (Edited by Walter Hooper.)
12. Lewis, C.S. 1970. *God in the Dock: Essays on Theology and Ethics*. Grand Rapids, MI: Eerdmans, p. 229. (Edited by Walter Hooper.)
13. Lewis, 1970, p. 229.

existed, otherwise nothing would exist now.

Lewis humorously but aptly explained the evolution myth as humans evolving from a weak, tiny spark of life that began amidst the huge hostilities of the inanimate, the product of another millionth millionth chance, adding that even 7th-graders understand the math involved in the millionth millionth chance.

Relating the evolution myth to a play, Lewis said it is a story "preceded by the most austere of all preludes: the infinite void, and matter restlessly moving to bring forth it knows not what."[14] He added that "the Myth gives us almost everything the imagination craves—irony, heroism, vastness, unity in multiplicity, and a tragic close. ... That is why those of us who feel that the Myth is already dead for us must not make the mistake of trying to 'debunk' it in the wrong way. ... It is our painful duty to wake the world from an enchantment."[15] He knew that the evolution myth "has great allies. Its friends are propaganda, party cries, and bilge, and Man's incorrigible mind."[16]

An example Lewis employed was how this satanic scheme is embraced by penning the demon dialogue in his book *The Screwtape Letters*: "So inveterate is their appetite for Heaven that our best method, at this stage, of attaching them to earth is to make them believe that the earth can be turned into Heaven at some future date by politics or eugenics or 'science.'"[17] As we are now seeing the fanaticism over evolutionism, we can understand the prophetic statement of Lewis when he said, "I dread government in the name of science. That is how tyr-

---

14. Lewis, C. S. 1947, 1980. *The Weight of Glory and Other Addresses, Revised Edition*. New York: Macmillan, p. 79.

15. Lewis, 1967, p. 116.

16. Lewis, 1967, p. 115.

17. Lewis, C.S. 1996. *The Screwtape Letters*. New York: HarperOne, p. 156.

annies come in."[18] And come in they have. C.S. Lewis, although not a perfect man, towards the end of his life was not an evolutionist in any form. *C.S. Lewis: Anti-Darwinist* provides ample ammunition to win that argument.

Dr. Jerry Bergman continues to muzzle the megalomaniacal 'True Believer in Evolutionism' as villains who could easily be compared to General Zod in DC Comics; always defeated, yet ever trying to win by any possible vile means. In fact, the DC series "Infinite Earths" is as believable as evolutionism on our unique Earth. That said, like the imaginary Zod, 'True Believers in Evolutionism' have caused a lot of havoc and harm in the world. Thank God for giving us an intellectual warrior—a superhero, Dr. Jerry Bergman—who can resist the archenemies of truth and all that is good.

---

18. Lewis, 1970, p. 315.

Acknowledgments

Marilyn Dauer, Grieg Davis, especially Bryce Gaudian and David Bassett, M.S. who edited the entire book. I also am grateful to the staff at Marion E. Wade Center containing the C.S. Lewis Collection at Wheaton College in Wheaton, Illinois, for their support on this project.

1

Introduction

THIS IS THE SECOND BOOK I have authored about Clive Staples Lewis, (C.S.) Lewis, and his views about science and Darwinism. Since my first book, I have received a large number of e-mails and much feedback, almost all very positive. I was also honored by several very positive reviews on Amazon, except for one which I will respond to later.

The main issue I am attempting to deal with in this book is, it is widely believed that C.S. Lewis was a theistic evolutionist, or at least was neutral, about Darwinism. Typical of this belief is Christiansen who claimed, "the theory of organic evolution caused him no problems."[1] He added, "Modern science claims that man descended from animals… Lewis has no objections to the scientific view." The fact is, Lewis was anything but neutral on this topic. Evolution was a concern for him even before he became a Christian, and permeated his writings until he died.

Furthermore, if "Lewis too willingly accepts the Darwinian thesis of man's ascent from the beasts, perhaps it is because he holds a high

---

1. Christensen, Michael. 1979. *C. S. Lewis on Scripture*. Waco, TX: Word Books, p. 31.

view of animals in general."[2] Then, Christensen adds, what "he [Lewis] adamantly objects to is the notion of 'emergent evolution'—that man is naturally progressing upwards toward perfection and [more] complete knowledge." These two claims about Lewis' views appear contradictory. This confusion will be dealt with in some depth in the following pages. Lewis was consistent in his use of terms, but those who write about Lewis were often confused, as it appears Christensen obviously was.

Regarding how effective Lewis' rhetoric was, one evolutionist concluded that Lewis' anti-evolution writings were "responsible for millions and millions of people who refuse to admit that they evolved from early hominids."[3] Furthermore, Schwartz concluded, Lewis' opposition to Darwinism

> dominates the Space Trilogy and ... the seemingly impassable conflict between Christian tradition and the evolutionary or "developmental" tendencies of modern thought... the deep-seated conceptual paradigm, well established by the time of Darwin's monumental *Origin of Species* (1859), which transferred the focal point of creation from a transcendent God to the progressive development of Man" by evolution.[4]

The conclusion that Lewis was a theistic evolutionist relies on a select few isolated passages in his early writings, specifically in his books *Mere Christianity* and *The Problem of Pain*. These passages are considered in detail in several chapters in this book. In general, I now realize that the sections where I and others indicated Lewis accepted some form of Darwinism were misinterpreted and are not evidence

---

2. Christensen, 1979, p. 32.
3. Miller, Ryder W. (editor.) 2003. *From Narnia to A Space Odyssey*. New York: iBooks, p. 19.
4. Schwartz, Sanford. 2009. *C. S. Lewis on the Final Frontier: Science and the Supernatural in the Space Trilogy*. New York: Oxford University Press, p. 6.

that Lewis ever capitulated to Darwinism.

Several readers noted that there exists a number of very clear statements by Lewis that supported my thesis, but which I neglected to include in my first book. These references have been added to this edition. In my further reading of Lewis' writings, I encountered much information that significantly strengthened the conclusions of my original book. It is important to interpret Lewis in the context of all of his writings, and his writings, taken as a whole, to understand what individual sections were attempting to convey.

In one example, Lewis discussed evolution to explain what the Darwinists believe, not to convey his own beliefs, or his agreement with their beliefs. A good example is, in discussing scientists' beliefs about the origin of the universe, which are quite different than the beliefs held today, Lewis prefaced his discussion on cosmological evolution with "scientists think" not with I [Lewis] think, or "believe," or "agree with."[5] Theistic evolutionists have repeatedly misread, or pull his statements out of their context, to support what they want to believe about Lewis in support of their own view.

What Could One More Book Add?

Much has been written about C.S. Lewis and to the question "What could one more book add?" I respond, "A lot." My first book is currently in over 200 WorldCat libraries alone, and has sold very well for a book about a man who died almost 60 years ago. This new work, and my previous book, are the only full-length works that focus specifically on Lewis' view on the topics of naturalism, evolution, creation, and scientism. In short, this new work makes a far stronger case that Lewis rejected Darwinism and also theistic evolution, especially toward the end of his life. It also strengthens the case that I made in my first book, namely that he accepted the basic creation worldview.

---

5. Lewis, C.S. 1996. *The Problem of Pain*. New York: HarperOne, p. 6.

Defining Terms

A major problem in understanding the writings of C.S. Lewis is that several terms need to be clarified at the onset. To achieve this, I will allow one of the most respected and widely published evolutionists to do this for me, namely Michael Ruse, the author, or co-author, of over 50 books. He documents the fact that "Darwinian thinking, in particular since the publication of the two great works" *On the Origin of Species* and *The Descent of Man* "has taken on the form and role of a religion. One in opposition to the world system, Christianity."[6]

Ruse does not infer that Darwinists claim to believe in the Christian God. Nor is he referring to evolution defined as the findings of laboratory genetic research, or field research, but, in his words, "Evolutionism or Darwinism" is the *worldview* that Darwin ushered in to replace the Christian worldview. He calls Darwin the man who "absolutely, totally and completely… changed our world…."[7] In many ways, we in the West have experienced a Darwinian revolution, namely a world before Darwin (BD) and a very different world after Darwin (AD). This was C.S. Lewis' main concern which was widely reflected throughout his writings.

This revolution began before Darwin published his books noted above. Examples include the writings of "God's greatest gift to the infidel, David Hume," who Ruse states "tore into the foundations of Christianity with a vigor that has yet to be equaled, even by those dubbed the new atheists… he was scathing about some of the traditional arguments for God's existence," namely the "argument from design" which Hume "subjected to detailed and withering critical discus-

---

6. Ruse, Michael. 2017. *Darwin as Religion*. New York: Oxford University Press, p. ix.

7. Ruse, 2017, p. v.

sion."[8] Hume laid the foundation that allowed Darwin to construct his atheist edifice which now dominates the scientific establishment.[9] The world Darwin shattered was Christianity, specifically the Christian worldview, by "breaking the hold of the Church" on society.[10]

Another forerunner of Darwinism was Progressivism, a political position heard everywhere today from Democratic presidential candidates. Ruse writes, "Evolution arrived on the back of the social doctrine of Progress. It [progressivism] was therefore at once plunged into the battle against religion." Because, if "evolution is to move forward, it must speak to both the social and the theological, in its own support" and in opposition to the Church.[11] Ruse documents the fact that evolution is "in opposition to religion, the Christian religion."[12]

The cover blurb that summarizes Ruse's book states that "there was indeed a revolution [in the 19th century] and that naturalist Charles Darwin was at the heart of it. However, contrary also to what many think, this revolution was not primarily scientific… but more religious or metaphysical, as people were taken from the secure world of the Christian faith into a darker, more hostile world of evolutionism."

Ruse quotes extensively from historians, poets, novelists and even philosophers, to document his point in 310 pages plus 13 pages of small-print references. Most ironic, Ruse notes, are the so-called Christian evolutionists who attempt to blend together these two very opposing ideas, namely Darwinism and Christianity.

Lewis was *not* one of these Christian evolutionists and, in fact, would agree with Ruse's statement that "evolutionary thinking became

---

8. Ruse, 2017, pp.13-14.
9. Ruse, 2017, p. 14.
10. Ruse, 2017, p. 14.
11 Ruse, 2017, p. 18.
12. Ruse, 2017, p. 36.

something more [than science]. It became a secular religion, in opposition to Christianity. In the second half of the nineteenth century and into the first part of the twentieth century, Darwinian evolutionary thinking... became a belief system countering and substituting for the Christian religion."[13] Lewis openly recognized this fact, and much of his writing was designed to oppose the Darwinian worldview.

In another book, Ruse documents the results of this new religion in his own life.[14] Raised a devout 'Christian,' at age 20, his faith faded in college, never to return. He was so enamored with Darwin in college that he became a philosopher of science specializing in Charles Darwin's theory of evolution.[15] Ruse's first book, *The Darwinian Revolution: Science Red in Tooth and Claw*, was on Darwinism, and 50 more books soon followed on the same theme.[16]

Ruse's *Meaning to Life* book, a summary of where his long journey into Darwinism led him to, is most revealing. In this book, Ruse discussed in detail those he calls "our opponents, Christians, often the more evangelical kind."[17] Ruse admits that evolutionists are in a war against Christians and brags that the evolutionists are winning, partly due to his own testimony in the most well-known American creation court case.[18] Ruse agreed with the view that Darwin's goal was to murder God, a goal that he (Ruse) was very successful in helping to achieve, writing that...

---

13. Ruse, 2017, p. 82.

14. Ruse, Michael. 2019. *A Meaning to Life*. New York: Oxford University Press.

15. Ruse, 2017, pp. ix-x.

16. Ruse, Michael. 1979. *The Darwinian Revolution: Science Red in Tooth and Claw*. Chicago: University of Chicago Press.

17. Ruse, 2019. P. 82.

18. *McLean v. Arkansas Board of Education*, 529 F. Supp. 1255 (E.D. Ark. 1982).

Darwin knew his theory was much better than [the rival evolution theory by] Chamber's ... but it was evolutionary and materialistic nonetheless... When telling Hooker of his evolutionism, Darwin confessed that it was like admitting to a murder. It was murder ... of Christianity, and Darwin was not keen to be cast in this role. Hence the *Essay* [which became the *Origin of Species*, finally published in 1859,] went unpublished.[19]

The result of rejecting Christianity, and replacing it with Darwin, was disastrous in Ruse's own life. His pessimistic conclusion was, "You are born. You live. Then you die. If you don't think so, then you should! We come from an eternity of oblivion. We return to an eternity of oblivion... In the end, you know truly that it [life] doesn't mean a thing."[20]

Ruse offers this view not only as his opinion, but as the supposed wisdom of past ages. He quotes celebrated Russian author Leo Tolstoy (1828-1910) who opined, "everything we do and say ends up as 'stench and worms.'" Tolstoy later became a devout Christian and aggressively rejected his previous view.[21] Ruse then quotes Harvard Professor William James (1842-1910), who declared, "We are all such helpless failures in the last resort."[22] He then jumps forward to the pessimist Albert Camus (1913-1960), who wrote that there "is but one truly serious philosophical question, and that is suicide."[23]

In the end, Ruse's attempts to determine whether Darwinism is a religion that will satisfy mankind's craving for justice, love, purpose, and meaning. He concludes, in agreement with Darwinism, that the

---

19. Ruse, 1979, p. 185.
20. Ruse, 2019, p. 1.
21. Ruse, 2019, p. 2.
22. Ruse, 2019, p. 3.
23. Ruse, 2019, p. 6.

Darwinian world "is a bleak world indeed."[24] As a committed Darwinist, Ruse attempts to rehabilitate his God, Darwin. On the last two pages of his book he claims, "I can give you a good Darwinian account of Meaning in terms of our evolved human nature. This is not a weak substitute. This is the real thing."[25] The real thing, he eloquently documents, are the fruits of his religion of Darwinism mentioned in his book *A Meaning to Life*: "In the end, you know truly that it [life] doesn't mean a thing."[26] Much of Lewis' writing and lecturing was to oppose this bleak and grim worldview as openly espoused by Michael Ruse.

Summary

Microevolution, or variation within the Genesis kinds, which Lewis called evolution, should be distinguished "from what may be called the universal evolutionism" or developmentalism by which he means

> the belief that the very formula of universal process is from imperfect to perfect, from small beginnings to great endings, from the rudimentary to the elaborate: the belief which makes people find it natural to think that morality springs from savage taboos, adult sentiment from infantile sexual maladjustments, thought from instinct, mind from matter, organic from inorganic, cosmos from chaos. This is perhaps the deepest habit of mind in the contemporary world.[27]

The Reference Problem

One problem with referencing C.S. Lewis' writings is that so many different editions, reprints and compilations exist that referencing his work can be very problematic. I counted over 50 different editions

---

24. Ruse, 2019, pp. 97, 133, 134.
25. Ruse, 2019, p. 169.
26. Ruse, 2019, p. 1.
27. Lewis, 1996f, p. 104.

of *Mere Christianity* in English alone and no doubt my count is incomplete. Furthermore, *Mere Christianity* has been reprinted in several compilations of Lewis' writings. Also, American, British and other editions exist that make locating some of the quotes referenced in writings about C.S. Lewis difficult. For this reason, in this work, the book title can be used to allow the reader to locate the quotes of concern in his or her own edition of Lewis' writings.

Darwin as an elderly man. In most of the pictures of Darwin that survived he does not appear to be a very happy man.

2
Lewis' Terminology for Darwinism

LEWIS CALLED DARWINISM "the great myth," "developmentalism," and "emergence theory." Developmentalism was once a common term for evolutionism. This book was written partly in response to critics who claimed that Lewis accepted macroevolution and would not agree, even partly, with either Intelligent Design or Creationism.[1] The key to understanding Lewis is that he believed

> 'a sane man accepts or rejects any statement, not because he wants to or does not want to, but because the evidence seems to him good or bad.' This statement encapsulates Lewis' approach to religion: **Follow the evidence**. The overarching project of Lewis' Christian writings is to *make the case that the evidence leads to Christianity.*[2]

---

1. For example, see Peterson, Michael L. 2010. "C. S. Lewis on Evolution and Intelligent Design." *Perspectives on Science and Christian Faith* **62**(4):253-266, December.

2. Wielenberg, Erik J. 2008. *God and the Reach of Reason: C. S. Lewis, David Hume, and Bertrand Russell.* New York: Cambridge University Press, p. 56. Emphasis added.

In short, "C.S. Lewis ... believed in argument, in disputation, and in the dialectic of Reason because he believed that **the main business of life was a bold hunt for truth**."[3] Lewis also feared that science would lead to "rejection of absolute standards and traditional, objective values" which are illustrated by Ruse's book *Darwinism as Religion*, leaving

> no defense against what some men might do with the powers of science. His [Lewis'] love of individual freedom and his appreciation for men as creatures of worth in God's sight caused him to fear what might be done to men if science, without the old values to restrain it, were to be given the power of government to enforce what a few men might plan for all the rest.[4]

This is exactly what has happened today in many areas, including the forced teaching of Darwinism in the schools and termination of teachers who cause students to question the validity of Darwinism.[5] In his book *The Abolition of Man*, Lewis described the debunking of traditional values that he observed were "in education, in sociology, psychology, biology, and philosophy" where students are

> taught that values are nothing but subjective reactions depending on body chemistry and environment. Then he [Lewis] traced the process of the "Conquest of Nature" by which one phenomenon after another is reduced to a quantitative object ... the mind, reason, and all human characteristics become "merely" objects in Nature when men study them as such, forgetting that there is mystery and reality which cannot be stud-

---

3. Murphy, Brian. 1983. *C.S. Lewis (Starmont Reader's Guide, 14)*. Mercer Island, WA: Starmont House, p. 11. Emphasis added.
4. Crowell, 1971, p. iii.
5. Bergman, Jerry. 2021. *Slaughter of the Dissidents: The Shocking Truth About Killing the Careers of Darwin Doubters*. Third Edition. Southworth, WA: Leafcutter Press.

ied empirically.[6]

This prediction by Lewis has also now been eloquently fulfilled in Western society. The claim that Lewis rejected a particular argument for God is also examined, namely, in Lewis scholar DePaw University President Erik Wielenberg's words, the...

> oldest, most popular, and most enduring types of theistic arguments on the market: the argument from design ... A key element of the argument from design is the idea that the observable universe has certain features that indicate intelligent design at work in its formation. This argument comes in many varieties and has had many defenders.[7]

Although Wielenberg claims that Lewis was "no friend of the design argument," on the contrary, we will show *the main apologetic argument that Lewis used to defend his worldview was the design argument.*

Since his death in 1963, many unpublished Lewis manuscripts and letters have been published, thanks to Lewis' dedicated private secretary, Walter Hooper.[8] These manuscripts, some that Lewis did not feel comfortable publishing in his lifetime, have... helped enormously in clarifying Lewis' conclusions, not only about Darwinism, but also about the secular science establishment. Many of Lewis' arguments have proved prophetic for today's world. His concerns have been borne out in striking ways especially during the last few decades in the Intelligent Design and creation-evolution struggles.

Lewis recognized that when a word has several meanings, it is more likely to mean its biological sense rather than its historical meaning.[9]

---

6. Crowell, 1971, pp. ii-iv.
7. Wielenberg, 2008, p. 57.
8. McGrath, Alister. 2013. *C. S. Lewis: A Life*. Carol Stream, IL: Tyndale, p. 367.
9. Lewis, C. S. 1967. *Studies in Words. Second edition.* New York: Cambridge University Press. pp. 12-13.

An example is that he used the word *evolution* to mean observable physical changes from generation to generation, not evolution from a primitive ape like ancestor to modern man. What he condemned was not observable biology that we see from generation to generation, but Darwinism (or Neo-Darwinism) from ape to modern man development. Lewis quotes George Bernard Shaw, who labeled Darwinian evolution as "the religion of the Twentieth Century" as Ruse defined it.[10] Shaw opined that Darwinian evolutionary thinking is "unmistakably the religion of the twentieth century, newly arisen from the ashes of pseudo-Christianity, of mere skepticism, and of the soulless affirmations and blind negations of the Mechanists and Neo-Darwinians."[11]

Myths About the Word Myth

One critic of my position, Amazon reviewer and ex-Young-Earth Creationist Michael C., wrote "C. S. Lewis clearly held that Adam and Eve were not spontaneously created, but came from a non-human anthropoid ancestor. He clearly held that there was death before the Fall. He clearly held that the early chapters of Genesis were more myth than history."[12] These common claims will be dealt with in detail in the following pages. For now, the term "myth" will be discussed because it often comes up in the following chapters, as well as in C.S. Lewis' writings.

A lot of myths exist about the word "myth," a word that Lewis often used. A common meaning of a myth is a widely held, but false,

---

10. Lewis, 1967 p. 300. The quote is from George Bernard Shaw *Back to Methuselah*. Lewis gave the page as iii but in my copy of Shaw, the Oxford University Press edition of 1947, it is on p. page lxviii.

11. Shaw, G. B. 1911 *Back to Methuselah*. London Constable. page lxviii.

12. https://amazon.com/productreviews/1532607733/ref=cm_cr_arp_d_viewpnt_rgt?ie=UTF8&filterByStar=critical&reviewerType=all_reviews&pageNumber=1#reviews-filter-bar. This review is grossly inaccurate, even claiming in 2018 that ICR was my employer! I was then teaching biology at a local college.

belief or idea. This meaning ignores the fact that the term "myth" is from the Greek word *mythos* which means "word in the sense of final authority."[13] The definition of myth used in the social sciences and philosophy, the fields of Lewis' expertise, is an account or narrative, "especially one concerning the early history of a people or explaining some natural or social phenomenon, and typically involving supernatural beings or events." In the words of the leading expert in the field of myths, the late Professor Levi-Strauss, the conclusion that creation myths were only naive attempts to explain reality is incorrect:

> Some claim that human societies merely express, through their mythology, fundamental feelings common to the whole of mankind, such as love, hate, or revenge, or that they try to provide some kind of explanations for phenomena which they cannot otherwise understand – astronomical, meteorological, and the like. But why should these societies do it in such elaborate and devious ways, when all of them are also acquainted with empirical explanations?[14]

In short, creation myths are accounts "concerning the early history of a people or explaining some natural or social phenomenon, and typically involving supernatural beings or events." As White states, to Lewis, "Myth represents ultimate, absolute reality."[15] For Lewis, "characteristics which separate man from the other forms of animal life [were] … grounded in the doctrine of Creation." For example, in one of Lewis' novels, Prince Caspian bemoaned that he wished he came

---

13. Hamilton, Virginia. 1988. *In the Beginning: Creation Stories from Around the World*. San Diego: Harcourt Brace Jovanovich, p. x.
14. Levi-Strauss, Claude. 1996. The Structural Study of Myth. In *The Continental Philosophy Reader*, pp. 305-327. Edited by Richard Kearney and Mara Rainwater. New York: Routledge, p. 308.
15. White, William Luther. 1969. *The Image of Man in C. S. Lewis*. Nashville, TN: Abingdon Press, p. 37.

from a more honorable lineage than he was told he came from.

Aslan's answer to this charge was that, no, you came from an honorable lineage, namely, "You come of the Lord Adam and the Lady Eve.... And that is both honor enough to erect the head of the poorest beggar, and shame enough to bow the shoulders of the greatest emperor on earth. Be content.... You were a son of Adam and came from the world of Adam's sons."[16] This tendency to teach his views will be covered in detail in the following pages. Lewis also occasionally used the term "myth" to mean a false statement believed to be true by many. I will attempt to show by the context which use Lewis employed in each specific situation.

---

16. Lewis, C.S. 2004. *The Chronicles of Narnia*. New York: HarperCollins, p. 416.

Lewis as a young man, age 17 when he was still a student.

3

Why C.S. Lewis is Important Today

OXFORD[1] UNIVERSITY PROFESSOR C. S. Lewis was widely regarded as one of the most important Christian apologists of the last century.[2] One evidence of this fact is the over 60 books and monographs that he authored have sold over 70 million copies (the majority remain in print), and seven journals are, or have been, devoted to Lewis.[3] He was regarded as "one of the intelligent giants of the twentieth century and arguably one of the most influential writers of his day."[4] He has, at least in much of his writing, managed to escape the Christian/religious label. One example of this was noted by Richard Corwin who commented that when growing up "I was able to read the writings of Lewis, when the mere appearance of a Bible in our home was sufficient

---

1. He taught at Cambridge since 1954.
2. Wielenberg, 2008.
3. Joshi, 2003, p. 105.
4. Corwin, Richard. 2016. *Creation Evolution and the Handicapped: Crushing the Death Image.* Bloomington, IN: WestBow Press division of Thomas Nelson, p. 40.

to provoke a wave of indignation."⁵

Lewis was also a major writer in the new science-fiction genre that began with Jules Verne (1828-1905) and, more importantly, with evolutionist H. G. Wells who Lewis would have a lot to say about.⁶ Toward the end of his long career, Lewis concluded that the modern theory of evolutionary naturalism, often called Darwinism in honor of the man who was one of the most important modern popularizers of the theory, was one of the most destructive ideas ever foisted on civilization.⁷

Lewis also called Darwinism, as defined by Ruse above, *emergent evolution* and Developmentalism to distinguish it from what we today refer to as microevolution or variation within the Genesis kinds as determined by empirical study. Lewis concluded that emergent evolution was "pure hallucination."⁸ Lewis explained that, for the practical experimental scientist, "Evolution is purely a biological theorem" which only attempts to explain certain minor changes in life or the environment, and

> makes no cosmic statements, no metaphysical statements, no eschatological statements [as does Darwinism]. Granted that we now have minds we can trust, granted that organic life came to exist, it tries to explain, say,

---

5. Corwin, 2016, p. 40.
6. Kilby, Clyde S. 1985. "Into the Land of Imagination." *Christian History* **4**(3):16-18.
7. Bergman, Jerry. 2017. *Evolution's Blunders, Frauds and Forgeries*. Atlanta, GA: CMI Publishing; 2014. *The Darwin Effect: Its Influence on Nazism, Eugenics, Racism, Communism, Capitalism & Sexism*. Green Forest, AR: Master Books; 2012. *Hitler and the Nazi Darwinian Worldview: How the Nazi Eugenic Crusade for a Superior Race Caused the Greatest Holocaust in World History*. Kitchener, Ontario, Canada: Joshua Press.
8. Lewis, C. S. 1962. *They Asked for a Paper: Papers and Addresses*. London: Geoffrey Bles, Ltd. This collection includes his 1954 presentation to Cambridge titled *De Descriptione Temporum*, p. 164; Lewis, C. S. 1996. *The Weight of Glory*. New York: Simon & Schuster, p. 138.

how a species that once had wings came to lose them. It explains this by the negative effect of environment operating on small variations. *It does not in itself explain the origin of organic life, nor of the variations*, nor does it discuss the origin and validity of reason.[9]

Lewis accepted the form of evolution that taxonomists often call microevolution, and creationists refer to as variation within the Genesis created kinds (aka *baramin*). This is obvious in a question he was asked by a Mr. J. W. Welch: "Could the Anvil [the name of the radio program] tell me how to set about reading the Bible. I know that the Genesis account [of creation] is untrue, and some Old Testament miracles are legendary. ... Well what are we to believe?" Lewis answered Welch: "I'd like to say straight away that if this man knows the Genesis account [of creation] to be untrue, I'm wondering where he's picked up his information because I don't know anything about that from human knowledge—I believe it to be true."[10]

Lewis used the very descriptive term "biolatry" to aptly describe what 'True Believers in Evolutionism' do when they make evolution into an intelligent force. Rhetorically, he asked, "Does the whole vast structure of modern naturalism depend, not on positive evidence but simply on an *a priori* metaphysical prejudice? Was it devised not to get the facts but to keep out God?"[11] Naturalism is the view that the entire universe and the natural world can be explained by natural law. When Napoleon asked French polymath Pierre-Simon Laplace where God fit into his scientific work, Laplace famously replied, "Sir, I have no need of that hypothesis"(i.e., God, because science explains everything).

Lewis correctly feared "what man might do to mankind" because of

---

9. Lewis, 1967, p. 107. This quote is part of the essay "The Funeral of a Great Myth." Emphasis added.
10 Phillips, Justin. 2002. *C. S. Lewis In A Time Of War*. New York: HarperSanFrancisco, pp. 309-310.
11. Lewis, 1980a, p. 136.

the view "that morality is relative and that moral standards have grown from mere impulses, from chemical reactions and responses which are in turn simply part of the irrational, blind development of organic life from the inorganic" as Darwinism teaches[12] Lewis cited Nazism as one example that supported his concern.

Evolutionary naturalism and macroevolution, as interpreted by leading scientists, such as those elected as members of the National Academy of Science, *does* make cosmic, metaphysical and eschatological statements. Microevolution (variation within the original Genesis kinds) limits itself to what biologists can study. For example, macroevolution would explain epigenetic modifications and the loss of biological organs or functions, such as typified by the blind cave fish. Lewis rejected the former, the macroevolution idea, and accepted the latter, the study of microevolution.

Some of the many reasons Lewis gave for rejecting evolutionary naturalism, many of which are detailed extensively in his writings, are summarized in the following chapters in this book. It is also clear that Lewis would be very supportive of the Intelligent Design movement today. We will also document that claims such as those by Professor Michael Peterson, that Lewis accepted "evolution as a highly confirmed scientific theory"[13] (given the orthodox definition of evolution used by most scientists today), are powerfully refuted by Lewis' own words.

Orthodox evolution is often called Darwinism to differentiate it from the term evolution, which often can loosely refer to anything from the differences between a son and a father, to scientism. The term Darwinism is often used instead of evolution or scientism to make this distinction clear. A major reason for this is because Darwin's theory

---

12. Crowell, Faye Ann. 1971. *The Theme of the Harmful Effects of Science in the Works of C. S. Lewis*. M.A. Thesis, Texas A & M University, p. 12.
13. Peterson, Michael. L. 2010. "C. S. Lewis on Evolution and Intelligent Design." *Perspectives on Science and Christian Faith* **62**(4):253-266, December, p. 253. See also http://biologos.org/author/michael-l-peterson.

required believing in *philosophical materialism*, the conviction that matter is the stuff of all existence and that all mental and spiritual phenomena are its by-products. Darwinian evolution was not only purposeless but also heartless—a process in which ... nature ruthlessly eliminates the unfit. Suddenly, humanity was reduced to just one more species in a world that cared nothing for us. The great human mind was no more than a mass of evolving neurons. Worst of all, there was no divine plan to guide us.[14]

Darwin "was keenly aware that *admitting any purposefulness* whatsoever to the question of the origin of species would put his theory of natural selection on a very slippery slope" that would eventually lead to its rejection.[15] Darwin himself believed that his theory of natural selection replaced God, and actually was a god: "I speak of natural selection as an active power or Deity... it is difficult to avoid personifying the word Nature; but I mean by nature, only the aggregate action and product of many natural laws."[16] Furthermore

> Darwin realized that it would weaken his whole argument if he permitted his account of evolution to stop short of the highest forms of intelligence. Once he admitted that God might have intervened in an act of special creation to make man's mind, others might argue. "In that case, why not also invoke the aid of God to explain the worm?"[17]

Darwin also clearly taught atheistic evolution, stressing "I would give absolutely nothing for the theory of natural selection if it requires miraculous additions at any one stage of descent" which would make

---

14. Levine, Joseph S. and Kenneth R. Miller. 1994. *Biology: Discovering Life,* Second edition. Lexington, MA: D.C. Heath, p. 161. Emphases in original.
15. Turner, J. Scott. 2007. *The Tinker's Accomplice: How Design Emerges From Life Itself.* Cambridge, MA: Harvard University Press, p. 206. Emphasis added.
16. Darwin, Charles. 1897. *The Origin of Species.* Vol. 1, Sixth edition. New York: D. Appleton, p. 99.
17. Gruber, Howard E. 1974. *Darwin on Man: A Psychological Study of Scientific Creativity,* Second edition. Chicago, IL: University of Chicago Press, p. 202.

"my Deity 'Natural Selection' superfluous" and would "hold *the* Deity—if such there be—accountable for phenomena which are rightly attributed only to" natural selection."[18]

Lewis' writings often effectively argued against natural selection, at least indirectly. In one example, he reasoned that a real human need must have a real object. For example, hunger is a real human need that has a real object, namely food. Sexual desire is a real human need that has a natural object, namely sexual satisfaction. The need for something beyond what is necessary for survival is also a real need, Lewis taught, which is met in music, art and other creative endeavors.

Most people also experience a longing that is not satisfied by material things, referred to by some as spiritual needs. Lewis noted that some people claim that this need can be explained on the "grounds of evolution by natural selection" but the "price one pays for taking this line is that it makes the desires in question unattainable in principle if our 'infinite longings' do not mean that we shall never be satisfied."[19] He explained this idea in more detail which we will cover later.

Lewis and Education

In a well-documented work on C. S. Lewis and education, the authors discuss Lewis' view that a non-teleological approach to education is insufficient, and "Lewis' work reminds us … that the substratum of institutions and all economy and social action is the human being." Also, an important part of human beings is "a teleology of SI [Social Institutions] from what man is and was designed to be."[20] The authors

---

18. Darwin's June 5, 1861 letter to Asa Gray; Darwin, 1897, p. 99; Darwin, 1896, Vol. 2, pp. 165-166; Darwin, Charles. 1991. *The Correspondence of Charles Darwin: 1858-1859* – Vol 7. Cambridge University Press, p. 345 (edited by Frederick Burkhardt.); Moore, James. 1979. *The Post Darwinian Controversies*. New York: Cambridge University Press, p. 322.
19. Purtill, Richard L. 2004. *C. S. Lewis' Case for the Christian Faith*. San Francisco, CA: Ignatius Press, pp. 18-19.
20. Loomis, Steven R. and Jacob P. Rodriguez. 2009. *C. S. Lewis: A Philosophy of*

recognize that "there is a strong materialist (naturalist) objection to the teleological view for what will be obvious reasons." Loomis concluded we derive our metaphysics from "two theories: the atomic theory of matter and the evolutionary theory of biology."[21] One

> of Darwin's greatest achievements was to drive teleology out of the account of the origin of species. On the Darwinian account, evolution occurs by way of blind, brute, natural forces. There is no intrinsic purpose whatever to the origin and survival of biological species. We can, arbitrarily, define the 'functions' of biological process relative to the survival of organisms, but the idea that any such assignment of function is a matter of the discovery of an intrinsic teleology in nature, and that functions are therefore intrinsic, is always subject to [speculation].[22]

Loomis and Rodriguez add that, as Lewis also recognized, Darwinism results in "a situation wherein the human being and the world… are reducible to mere Nature and its processes of cause and effect."[23] Before the details of Lewis' reasoning are reviewed, some background on Lewis is necessary.

Education. New York: Palgrave Macmillan, p. 42.

21. Loomis and Rodriguez, 2009, p. 42.
22. Loomis and Rodriguez, 2009, p. 42.
23. Loomis and Rodriguez, 2009, p. 44.

A book by, and about, Warren Lewis,
Copy of the book cover.

4

What Made Lewis the Man He Became

CLIVE STAPLES LEWIS was born in Belfast, Ireland on Nov. 29, 1898 to Albert James Lewis, a police court solicitor (a type of lawyer), and Florence (Flora) Augusta Hamilton, a mathematician with a First Class B.A. Honors Degree in logic, and a Second Class Honors in mathematics from Queens College in Belfast, Ireland.[1] She was the daughter of a clergyman, and her family includes many generations of clergymen and lawyers.[2]

His father, Albert, was educated at Lurgan College and enjoyed a good position with the government. The upper middle class Lewis family resided in a large home in a good neighborhood. Both parents were "bookish" and spent much time reading in this pre-smart-phone era. In fact, "Albert and Flora shared only one leisure activity, voracious reading," often spending all evening after dinner until bedtime reading.[3] They read a large variety of books, including the classics and

---

1. Dorset, Lyle. 1985. "C. S. Lewis: A Profile of His Life." *Christian History* **4**(3):6-11.
2. Lewis, C. S. 2002. *Surprised by Joy*. New York: Barnes & Noble, p. 1.
3. Schultz, Jeffrey and John G. West. 1995. *C. S. Lewis Readers' Encyclopedia*. Grand

academic books on history, literature and science. Consequently, the home Lewis was reared in had

> books in the study, books in the drawing room, books in the cloakroom, books (two deep) in the great bookcase on the landing, books in a bedroom, books piled as high as my shoulder in the cistern attic, books of all kinds reflecting every transient stage of my parents' interest, books readable and unreadable, books suitable for a child and books most emphatically not. Nothing was forbidden me. In the seemingly endless rainy afternoons I took volume after volume from the shelves. I had always the same certainty of finding a book that was new to me as a man who walks into a field has of finding a new blade of grass.[4]

Both C. S. and his brother Warren followed their parents' example throughout their lives. Lewis' favorite reading included animals, detective books by Conan Doyle, and even science fiction, including that by H. G. Wells.[5] Reading led to Lewis' strong drive to write. As a professor in his department at Oxford until 1954, he spent eight to ten hours a day tutoring one or two students per session, and was assigned related academic activities, yet wrote over 60 books, an enormous amount for a full-time professor.

When he moved to Cambridge in 1954, Lewis had more time to write, resulting in a significant increase in productivity. He wrote, and/or answered, close to 10,000 letters in his lifetime, many of which were reprinted in a three-volume collection totaling over 3,200 pages of small print covering the dates 1905 to his death in 1963.[6] Unfortunately, he did not save many of the letters sent to him, though he faithfully answered, as far as is known, all of his mail, usually in the

---

Rapids, MI: Zondervan, p. 15.

4. Lewis, 2002, pp. 7-8.
5. Schultz and West, 1995, p. 15.
6. *The Collected Letters of C. S. Lewis* (2004). Grand Rapids, MI: Zondervan. (Edited by Walter Hooper.)

morning hours. Lewis felt that, as an author, he had an obligation to respond to his readers and supporters, and consequently was contentious in fulfilling this task.

His Brother Warren

Lewis had one brother, Warren Hamilton, nicknamed Witt, three years his senior. C.S.'s lifelong nickname was Jack. Witt was Lewis' lifelong companion until Lewis died, excepting only when they were separated by their time in military service.[7] They were allies and did most everything together, even creating their own made-up country which they named Boxen. In this imaginary country they created many individual inhabitants as well as the nation's four-hundred-year history.

The brother's close relationship is illustrated by an event that occurred when, as young boys, they were sitting at a table working on separate drawings. All of a sudden C. S. bumped his brother Witt's elbow accidentally, causing him to make an unintended line across his drawing. The matter was amicably settled by C. S. allowing Witt to make a comparable line on his (Lewis') drawing.[8] This is a good example of their give and take that resulted in their lifelong amiable and supportive relationship.

When C. S. Lewis was only four years old, his beloved dog, Jacksie, was hit by a car and died. Distraught, he told everybody that he changed his own name to "Jacksie," and, as a result, was called Jack for the rest of his life.

Lewis' mother, Flora, died of cancer on Albert's 45th birthday after 14 years of marriage to Jack's father. While his mother was dying, Lewis, who was nine then, fervently prayed for her health's return. When she died, he was very disappointed in God for not answering his sin-

---

7. Gibb, Jocelyn. 1965. *Light on C. S. Lewis*. London: Geoffrey Bles, pp. v. ii. [p. v. before p. ii.]
8. Lewis, 1996d, p. 91.

cere, fervent prayers.[9] He was taken into the bedroom where she lay to see her for one last time. Lewis described the experience by explaining he reacted with horror at seeing his mother's now lifeless body.

For many reasons, including he was only a child of nine when she died, Lewis had a very difficult time dealing with his mother's loss. This event reveals Lewis' sensitivity to pain and death that would again be reflected later when Lewis' wife, Joy, suffered from cancer.[10]

Albert Lewis never stopped grieving over the loss of his wife, and never remarried.[11] His life was his work. He was also a faithful member of St. Mark's Church until his own death, also of cancer, in 1929.

Shortly after his mother died, Lewis was sent off to boarding school. In Lewis' mind, he lost not only his mother, but also his father, a bond that was never fully mended. Nonetheless, his father fully supported Lewis' education and, without such support, he would never have become an Oxford professor and a world-renowned author. His brother, Warren, became his family, and they were very close for the rest of their lives.

From this background came the C. S. Lewis legend. He read widely as an adult, giving him a knowledge of literature that made him much sought-after for his company and conversation. Lewis enjoyed talking about literature, poetry, and religion into the late hours in college rooms for the rest of his life. His reading also included books that treated religion as "sheer illusion." Nor did C. S. at this time "hear or read any grounds for believing Christianity to be true."[12] It is not at all surprising that he became an atheist.

---

9. Peterson, Michael L. 2020. *C. S. Lewis and the Christian Worldview*. New York: Oxford University Press, p. 125.
10. Lewis, 2002, p. 17.
11. Lewis, C. S. 1991. *All My Road Before Me: The Diary of C. S. Lewis, 1922-1927*. New York: Harcourt Brace Jovanovich, p. 467. (Edited by Walter Hooper.)
12. Schultz and West, 1995, p. 20.

Marriage

Lewis was single for most of his life, marrying only at age 57 in 1956. He married the daughter of a New York Jewish couple, Joy Davidman Gresham, to enable her to remain in England after her visa expired. Often referred to as a child prodigy and a very talented writer in her own right, she was a graduate of Hunter College, and earned a master's degree from Columbia University in English literature at age twenty in 1935. She was married twice. Her first husband suffered a heart attack, causing her to think about religious matters, including the study of, among other writers, C. S. Lewis, who was critical in her conversion from Judaism to Christianity. She also then began attending a Christian church. Lewis met her when she and her two boys, Douglas Howard Gresham and David Gresham, came to visit him in London after she had been writing to him for some time.

Lewis' marriage to Joy, at first for convenience, soon became the love of his life, lasting over three years until she died of cancer. Lewis wrote an entire book, *A Grief Observed*, about his relationship with his wife.[13] Writing was clearly very therapeutic for him, as it is for many authors.

Lewis knew she had cancer when he married her. In 1957, after the laying on of hands and prayer by a minister, she made a recovery which even the doctors claimed was miraculous.[14] Both she and Lewis knew this hiatus may not be permanent, but felt God was giving them more time to develop their new marriage. After two years, the cancer came back in full force and spread throughout her bones. Three years after they legitimately married, she died at his home in Headington, Oxford, England. Her two boys were then reared by Lewis.

---

13. Lewis, 1996a.
14. Dorsett, Lyle W. 1983. *And God Came In*. New York: Macmillan. [The definitive biography of Helen Joy Davidman (1915-1960), the wife of C.S. Lewis from 1956/1957-1960.]

His Death

C. S. Lewis died at the young age of 64 of kidney failure on Nov. 22, 1963, the same day John F. Kennedy was assassinated and the well-known Darwinist proselytizer, Aldous Huxley, died.[15] C.S. Lewis' lifelong smoking habit likely was a major contributor to his premature death.

Over a half-century after his death, his books still sell several million copies annually.[16] A survey of 101 church leaders found that Lewis' book *Mere Christianity* was on the list of the top ten most influential books they had ever read.[17] Lewis was even called "one of the most brilliant minds of the 20th century."[18] Lewis has also "succeeded as few others in causing Christianity to be discussed seriously and publicly."[19]

Lewis was then, and still is now, one of the most celebrated Christian apologists of the last century. One would never have predicted this level of success from his early experience of publishing. Prior to his conversion, Lewis published three books, including two on poetry, *Spirits of Bondage* and *Dymer*. Neither one were sales successes, and most of the reviews were poor.[20]

---

15. Joshi, 2003, p. 109.
16. Yancey, Philip. 2010. *What Good is God?* New York: Faith Words, p. 100.
17. Rainer, Tom S. 2001. *Surprising Insights from the Unchurched and Proven Ways to Reach Them.* Grand Rapids, MI: Zondervan, p. 149.
18. Crowell, 1971, p. 4.
19. Deasy, Philip. 1958, "God, Space, and C. S. Lewis." *Commonweal* **68**(16):421-425, July 25, p. 421.
20. Burson, Scott and Jerry Walls. 1998. *C. S. Lewis and Francis Schaeffer: Lessons for a New Century from the Most Influential Apologists of Our Time.* Downers Grove, IL: InterVarsity Press, p. 31.

Part of Oxford University. One of the Oldest Universities in the World.

5

Lewis' Educational Background

LEWIS EARNED A triple first-class degree in philosophy from Oxford University. He was also fluent in six languages.[1] His mother began teaching him Latin and French when he was around age eight, and he later added Greek, Italian, and German to his study of languages.[2] Lewis was schooled by private tutors until age nine when his mother died from cancer in 1908, three months before Jack's tenth birthday. The family never fully recovered from her death.

Both boys felt estranged from their father, and home life was never again warm and satisfying as it was when their mother was alive. His father even sent him to live and study at Wynyard School in Watford, Hertfordshire, a bad experience under a mentally unbalanced headmaster. After stints at other schools, Lewis finally entered Oxford University and joined the Officers' Training Corps. At age 18 he was forced to leave Oxford during WWI to serve in the military. He ended up in the frontline trenches of the 3rd Somerset Light Infantry. After Jack

---

1. Crowell, 1971, p. 4.
2. White, 1969, p. 25.

was injured, he was sent home in May of 1918.[3] Lewis suffered from depression and homesickness during his convalescence.

Soon after his recovery in October, he was assigned to duty in Andover, England. His appearance was characteristic of the stereotypical eccentric genius: "his trousers were usually in dire need of pressing, his jackets threadbare and blemished by snags and food spots, and his shoes scuffed and worn at the heels."[4] Lewis eventually became a professor of literature at Oxford and taught both Philosophy and English, areas where language skills are critical.[5] He eventually published 125 essays and pamphlets, mostly on literary criticism and Christian apologetics.[6]

Lewis was an enormously influential Christian apologist in his day, and his popularity shows no sign of diminishing in ours.[7] Lewis' 60 books and monographs have been translated into 30 languages and have sold over 40 million copies. As of this writing, Lewis is the best-selling Christian author of all time.[8] Ryder W. Miller, in his study of C. S. Lewis, observed that he "had helped guide Christianity through the mid-twentieth century." His 29 British Broadcasting Corporation (BBC) lectures given during World War II reached an au-

---

3. Miller, 2003, p. 12.
4. C. S. Lewis scholar, author, and apologist. https://www.christianitytoday.com/history/people/musiciansartistsandwriters/cs-lewis.html.
5. Miller, 2003, pp.12-19.
6. Tolson, J. 2005. "God's Storyteller: The Curious Life and Prodigious Influence of C. S. Lewis, the Man Behind *The Chronicles of Narnia*." *U.S. News & World Report* **139**(22): 46-52, December 12, p. 48; Reid, Daniel G. 1990. *Dictionary of Christianity in America*. Downers Grove, IL: InterVarsity Press, p. 645.
7. Cunningham, Richard B. 2008. *C. S. Lewis: Defender of the Faith*, Wipf and Stock, p. 14; Green, Roger Lancelyn. 1963. *C. S. Lewis*. London: Bodley Head.; Green, Roger Lancelyn and Walter Hooper. 1974. *C. S. Lewis: A Biography*. New York: Harcourt Brace & Company. Revised edition, 1994. New York: Harvest Books..
8. Reid, 1990, p. 645.

dience estimated at 600,000, an enormous number back then.[9] And, Miller opined, the chances of finding another writer with the talent and Lewis' position in academia were slim to none.[10]

Lewis wrote about a remarkably wide range of subjects.[11] He penned nine books and about 30 essays that explored science and its impact on our modern culture.[12] Lewis was also very concerned about the "watering down of Christian doctrine by other theologians, especially the modernists" that was common in his day.[13] All total, he produced 49 major titles from 1919 to 1967, including some published posthumously, an output level that, as noted, was "astonishingly large" for a full-time professor.[14] These include the last book Lewis wrote, titled *The Discarded Image,* that

> critically examined the nature of scientific revolutions, especially the Darwinian revolution in biology. Lewis' personal library, meanwhile, contained more than three dozen books and pamphlets on scientific subjects, many of them dealing with the topic of evolution. Several of these books were marked up with underlining and annotations, including Lewis' copy of Charles Darwin's *Autobiography*.[15]

Lewis eventually chose science fiction as the vehicle for reaching the masses of his unbelieving countrymen, but was soon accused of smuggling Christian values into the science fiction of his famous space

---

9. Walsh, Chad. 1949. *C. S. Lewis: Apostle to the Skeptics*. New York: Macmillan, p. 9.
10. Miller, 2003, p. 11.
11. Bredvold, L. 1968. The Achievement of C. S. Lewis, *The Intercollegiate Review* 4(2–3):1–7, p. 1.
12. West, 2012, p. 11.
13. Joshi, 2003, p. 109.
14. White, 1969, p. 21.
15. West, 2012, p. 11.

trilogy and other fiction books.[16] In spite of, or because of these factors, Lewis influenced the works of some of the greatest science fiction writers of the last century, including Ray Bradbury, George Lucas (of Star Wars fame), Philip K. Dick and Arthur C. Clarke.

His Christian-based fiction book *The Lion, the Witch and the Wardrobe*, which was part of his seven-volume *Chronicles of Narnia* series, was adapted for a major Hollywood film.[17] Lewis' *Mere Christianity, The Problem of Pain, The Screwtape Letters,* and *The Abolition of Man* are all still extremely popular and influential Christian apologetic works. In April of 2000, *Mere Christianity* was voted by *Christianity Today* as the best Christian book published in the twentieth-century.

Lewis' science fiction created "a vision of the cosmos *without* evil extraterrestrials and a controlling evolutionary framework."[18] Lewis, as Peterson documents, effectively rebutted the arguments against Christianity posed by secular science. The problem was not the scientific method, nor empirical discoveries achieved by the scientific method, but by worldview conflicts posed by evolutionary materialism.[19] Furthermore, "Much of Lewis' devastating critique of scientism applied to evolutionism as well," a worldview which Lewis and others called *Developmentalism*.[20] Developmentalism, defined by Lewis scholars Loomis and Rodriguez, is "the idea that everything everywhere is improving" as the nature of things.[21]

---

16. Miller, 2003, pp. 12, 14.
17. Tolson, 2005, p. 46.
18. Miller, 2003, p. 14.
19. Peterson, 2020, p. 130.
20. Peterson, 2020, p. 132.
21. Loomis and Rodriguez, 2009, p. 79.

His Profession as a Professor

C. S. Lewis was a professor for almost 40 years, from 1925 until his premature death at age 64 in 1963.[22] From October 1924 until May of 1925, he taught philosophy at University College until a position opened at (St. Mary) Magdalen College, Oxford (founded in 1458) for which he applied and was hired.[23] He was soon a professor of medieval and early modern English literature at Oxford University, then later at Cambridge University, a topic covered below. He also was one of the "few dons whose lectures were always filled to overflowing."[24] As the stereotypical professor, Lewis once said that it is "important to acquire early in life the power of reading ... wherever you happen to be."[25]

Lewis Faces Religious Discrimination

In spite of the fact that he was the author of several highly acclaimed scholarly works of literary history, and that his lectures were among the most widely attended at Oxford University, Oxford never promoted him to Professor. As a result,

> Magdalen's brightest star was forced to slave away at time-consuming tutorials while many of his less brilliant colleagues were expected only to give lectures, thus freeing them up for research and publication. Still, this did not deter Lewis from producing a steady stream of popular and academic works as well as giving public lectures at numerous venues.[26]

Why he was not promoted was likely because "Lewis' refusal to hide his Christian faith under a bushel [which] hurt his chances for

---

22. Tolson, 2005, p. 46.
23. Duriez, 2013, p. 102.
24. Crowell, 1971, p. 4.
25. Lewis, 2002, pp. 53-54.
26. Markos, Louis, and David Diener. 2015. *C. S. Lewis: An Apologist For Education.* Camp Hill, PA: Classical Academic Press, p. 20.

promotion."²⁷ The problem was that "Lewis was unaware of quite how unpopular he was [due to his religious conclusions] in the English Faculty at Oxford, and indeed in the University at large."²⁸ Consequently, there was

> no chance of his being elected to the Merton Chair even though *The Allegory of Love* and *A Preface to 'Paradise Lost'* (quite apart from other lectures and learned articles which he had written to date) were far more interesting and distinguished than anything which his rivals for the job had produced. They, however, were safe men, worthy dullards: and that is usually the sort of man that dons will promote.²⁹

Another important factor was because

> Lewis had committed the unpardonable sin of being popular and reaching out to nonacademic readers. To make matters worse, he had spoken and published on subjects (like theology) that were outside his discipline. To be a generalist, and a popular one at that, was to fall foul of the rules of the club (dare I say inner ring?). Despite the advocacy of Tolkien, Lewis was unable to secure a professorship at Oxford.³⁰

Yet another reason for the discrimination he faced could be because "Oxford is a strange place, and dons are strange people. Brilliance in a colleague is quite as likely to excite their envy as their esteem, and, where mediocrity is the norm, it is not long before mediocrity becomes the ideal."³¹ For these reasons, the coveted Merton Chair of English Literature eluded Lewis until 1947 when the current

---

27. Markos, and Diener, 2015, p. 20.
28. Wilson, 1990, p. 208.
29. Wilson, 1990, p. 208.
30. Markos, and Diener, 2015, p. 20.
31. Wilson, 1990, p. 208.

occupant, David Nicholl Smith, retired, and Lewis assumed that he would be at least eligible, if not the likeliest candidate. He was weary of the repetitive round of undergraduate tutorials, and he disliked his colleagues at Magdalen. But even apart from these negative considerations, he considered himself worthy of the job. He was a popular and distinguished lecturer; and the job of an Oxford professor consisted very largely of lecturing in those days.[32]

In the end, it was not Lewis but his old tutor, F. P. Wilson, who was awarded the Merton Chair in 1947.[33] Wilson was a

Shakespearean scholar who could be relied upon not to cause trouble, and not to embarrass his colleagues by writing books about Christianity [as Lewis did]. Even colleagues who were Christians found Lewis' career as a popularizer embarrassing.... And they noticed that his variety of Christianity did not extend to meekness, or even necessarily to politeness.[34]

Fortunately, just

when Lewis had all but given up hope, Cambridge University came to the rescue and offered Lewis a professorship that they had created specifically to honor his work: a chair of medieval and Renaissance literature ... Lewis accepted the chair and held it from 1954 until his death in 1963. (Ironically, the college he worked for at Cambridge was [also] called Magdalene.)[35]

Also, ironically, his love for Oxford was so great that on weekends he lived at his Oxford home in the Kilns, the site of a former brickworks.

---

32. Wilson, 1990, p. 208.
33. Wilson, 1990, p. 208.
34. Wilson, 1990, p. 208.
35. Markos, and Diener, 2015, p. 20.

Lewis, A Respected Oxford Scholar

Lewis was not only a professor and historian, but frequently published in both widely respected academic as well as non-academic venues, the so-called popular press.[36] His doctoral student, Alastair Fowler, observed that Lewis' "scholarly and religious lives were really no more separable than two sides of the same coin."[37] Lewis existed in a hostile academic environment while at Oxford that was "resentful of his outspoken faith" and "denied him a full professorship... an injustice that rival Cambridge University later rectified," as noted by appointing Lewis as a professor.[38] The few colleagues that he had an excellent relationship with included a fellow Christian, J. R. R. Tolkien.[39] Often, though, while still at Oxford, Lewis "had to defend himself against Oxford's anti-Christian orthodoxy. One of these [secular] 'humanists'" he had to battle was British classical scholar Heathcote William Garrod (1878 –1960).[40]

His Mission to Bring Christianity to the Post-Christian World

Lewis saw his mission as bringing a viable, vibrant Christianity to the post-Christian world by his many lectures and books, a goal that he has achieved beyond anyone's expectation, most of all his own.[41] Lewis' focus was "on the core of Christianity that was common to all tradition-

---

36. Hart, Dabney. 1985. "Teacher, Historian, Critic, Apologist." *Christian History* 4(3):21-24.
37. Poe, Harry Lee and Rebecca Whitten Poe, (editors). 2006. *C. S. Lewis Remembered: Collected Reflections of Students, Friends & Colleagues.* Grand Rapids, MI: Zondervan, p. 126.
38. Yancey, 2010, p. 95.
39. Glyer, Diana Pavlac. 2007. *The Company They Keep: C. S. Lewis and J. R. R. Tolkien as Writers in Community.* Kent, OH: The Kent State University Press.
40. Fowler, Alastair. 2003. C. S. Lewis: Supervisor. *Yale Review* 91(4):64-80, October 1.
41. Tolson, 2005, p. 50.

al Christian denominations" which he called "mere Christianity."[42] When Lewis was writing

> the heart of traditional Christianity had lain in the historic Incarnation, the belief that God became Man, but by the time Lewis was converted to Christianity in 1931 Agnosticism was everywhere. The reliability of biblical scholarship had been punctured a good many years before by a number of things, amongst which were those 'prophets of enlightenment': Charles Darwin, Karl Marx, Friedrich Nietzsche, Sigmund Freud and Emile Durkheim—all of whose works Lewis probably was introduced to by his admired atheist teacher, Mr. Kirkpatrick].[43]

Many persons who later became well-known Christians were converted from atheism as a result of the influence of Lewis' writings. These include former atheist Chuck Colson and leading genetic scientist Francis Collins, previously Head of the Human Genome project and was until recently the Director of the National Science Foundation.[44] The popular author Philip Yancey also acknowledged the critical influence of Lewis in his life. Yancey writes that

> Lewis taught me a style of approach that I try to follow in my own writing. Most of us rarely accept a logical argument unless it fits our sense of reality, and the persuasive writer must cultivate that intuitive sense— much as Lewis did for me with his space trilogy before I encountered his apologetics. I came to believe in the invisible world only after tracing its clues in the visible world. Lewis himself converted to Christianity only after sensing that it corresponded to his deepest longings, his Sehnsucht [his "yearning" for more].[45]

---

42. Collins, C. John. 2011. *Did Adam and Eve Really Exist? Who They Were and Why You Should Care.* Wheaton, IL: Crossway, p. 13.
43. Hooper, Walter. 1996. p. 23.
44. Collins, 2011, p. 21.
45. Yancey, 2010, p. 94.

Yancey's statement that Lewis "came to believe in the invisible world only after tracing its clues in the visible world" is significant in view of the focus of this book. Likewise, many other modern apologists "cut their eye teeth on Lewis' works."[46] So many examples exist that Andrew Lazo and Mary Anne Phemister wrote, "Arguably no author of the twentieth century has had a greater spiritual impact on more people than Lewis."[47]

Lazo and Phemister collected the accounts of 55 writers, professors and others who were heavily influenced by Lewis, which they included in their book titled *Mere Christians: Inspiring Stories of Encounters with C. S. Lewis*. These notable persons include George Gallup Jr., Anne Rice, leading Christian author Randy Alcorn, Liz Curtis Higgs, Walter Hooper, Thomas Howard, Jill Briscoe and David Lyle Jeffrey.

A very notable example of a person impacted by Lewis was the poet and novelist Helen Joy Davidman Gresham (1915-1960) who became his wife. When Lewis was in his early fifties, he struck up a correspondence with her. Joy was a "nonpracticing Jew from the Bronx," New York, who had "been both an atheist and a communist."[48] Her

> reading of Lewis' apologetical works had helped lead her to faith in Christ. She visited Lewis in Oxford in 1952, and their friendship blossomed. Two years later, Joy was back in England with her two sons; she was now divorced from her alcoholic, serially unfaithful husband and was getting by as best she could.[49]

---

46. Markos, Louis. 2003. *Lewis Agonistes: How C. S. Lewis Can Train Us to Wrestle with the Modern and Postmodern World.* Nashville, TN: Broadman & Holman Publishers, p. 42.
47. Promotion for the book by Andrew Lazo and Mary Anne Phemister, (editors). 2009. *Mere Christians: Inspiring Stories of Encounters with C. S. Lewis.* Grand Rapids, MI: Baker Books.
48. Markos and Diener, 2015, p. 21. Reprinted in Lewis, 1962, *They Asked for a Paper.*
49. Markos and Diener, 2015, p. 21. Reprinted in Lewis, 1962, *They Asked for a Paper*, p. 14.

As a young woman, Joy had "assumed that science also proved that God didn't exist"[50] but traveled down a similar road Lewis did, and eventually became a believer. Also like Lewis, she eventually "realized she had believed in science the way religious people believe in God [when she realized it was] illogical, and, even worse, naïve to believe that the wonders of the atom had evolved blindly out of chaos [as evolution teaches]."[51]

She also concluded that the false gods of our age were "Sex, the State, Science, and Society," with "self" being the greatest god of all.[52] Thus, they had much in common and became soul mates, marrying in April of 1956. She turned out to be an important source of inspiration for both Lewis and his writing, and was the subject of his book *A Grief Observed*. As noted, she tragically died of cancer at the young age of 45.

His Concern about Evolution Surfaces

Both in his reading and formal education Lewis was exposed to Darwinism. As Corwin observed, early in his schooling and life, "there is ample evidence to show that Lewis had problems with scientific evolution and on a number of occasions attacked the myth head on."[53] This fact was commented on in some detail by the atheist philosopher S. T. Joshi, who arrogantly noted in response to the fact that "the great majority of his [Lewis'] readers, [are] manifestly naïve and untutored in philosophy, do not perceive their unsoundness—that an analysis of them becomes necessary." [54] Joshi's analysis is, essentially, if Lewis and his readers accepted what he calls the fact of evolution, this would ne-

---

50. Santamaria, Abigail. 2015. *Joy: Poet, Seeker, and the Women Who Captivated C. S. Lewis.* Boston: Houghton Mifflin Harcourt, p. 22.
51. Santamaria, 2015, p. 175.
52. Santamaria, 2015, p. 219.
53. Corwin, 2016, pp. 41-43.
54. Joshi, 2003, pp. 109-110.

gate Lewis' entire apologetic. Joshi's analysis will be covered in detail in the following chapters.

The Searcher. A C.S. Lewis centenary stature by Ross Wilson, located in Belfast.

6

Lewis Enters and Leaves Atheism

ALTHOUGH REARED A nominal Christian, Lewis became an atheist at age 15 due in part to his life experiences and his reading. These life experiences included the death of his mother when he was still a young boy. Another important influence was the body of conjecture that he learned in college against the main evidence for God, namely the cosmological (universal causality/design) argument. While at boarding school, Lewis often spoke out vehemently against religion. He writes that this conversion to atheism occurred at Chartres, the name he gave for the Cherbourg School he attended from January 1911 to June 1913. It was there that he made his

> first real friends. But there, too, something far more important happened to me: I ceased to be a Christian. The chronology of this disaster is a little vague, but I know for certain that it had not begun when I went there and that the process was complete very shortly after I left. I will try to set down what I know of the conscious causes and what I suspect of the unconscious.[1]

---

1. Lewis, 2002, p. 55.

An important reason he gave for accepting atheism was his "steadily growing doubts about Christianity."[2] The staunch so-called "rationalist" teacher and headmaster William T. Kirkpatrick (1848-1921) was one of several persons who influenced Lewis toward atheism. Kirkpatrick was originally headed for the Presbyterian ministry, but after three years of theological training became a disillusioned ex-Calvinist who attempted to convey his anti-Christian worldview to his students.

Lewis was also influenced by the then popular claim that life was poorly designed—a view now called dysteleology—made by certain evolutionists, such as the ancient Greek philosopher Lucretius. These claims have now been fully refuted.[3] Lewis wrote in his autobiography that "the atheism of his early years on the faculty at Oxford University" was due to science. Specifically, he judged that his atheism…

> was inevitably based on what I believed to be the findings of the sciences; and those findings, not being a scientist, I had to take on trust, in fact, on authority." What Lewis is saying is that somebody told him that science had disproved God; and he believed it, even though he knew [little] … about science.[4]

His study of natural science convinced him that life on the Earth was only a random accidental occurrence in our vast empty universe, and religion, as he learned from his study of the social sciences, was only the expression of a psychological need and of cultural values.[5] Lewis writes that the poor-design argument was an important reason for his atheism and that his…

---

2. Lewis, 2002, p. 72.
3. Bergman, Jerry. 2019. *The "Poor Design" Argument Against Intelligent Design Falsified*. Tulsa, OK: Bartlett Publishing.
4. Hooper, Walter (editor). 1996. *C. S. Lewis: A Companion & Guide*. London: HarperCollins Publishers, p. 23.
5. Peterson, 2020, p. 125.

> early reading—not only [H. G.] Wells but Sir Robert Ball—had lodged very firmly in my imagination the vastness of cold and space, the littleness of Man. It is not strange that I should feel the universe to be a menacing and unfriendly place. Several years before I read Lucretius I felt the force of his argument (and it is surely the strongest of all) for atheism ... *Had God designed the world, it would not be a world so frail and faulty as we see.*[6]

He then combined this directly atheistic thought with the great "'Argument from Undesign' ... that both made against Christianity. And so, little by little ... I became an apostate, dropping my faith with no sense of loss but with the greatest relief."[7] These experiences were very important in his later opposition to Darwinism. As described by Burson and Walls, having...

> been weaned on the science fiction of H. G. Wells, Lewis set out to produce an alternative version of the cosmos. Instead of sinister extraterrestrials and an overarching evolutionary framework... his science fiction painted a colorful and concrete picture of other worlds enchanted by a rich variety of rational, moral and aesthetically sensitive creatures."[8]

In short, Lewis spent his life responding to Wells' evolutionary worldview with the alternative — the creation worldview. He had to be careful in directly attacking evolution with biological arguments because he had very little formal education in this area. For that reason, he wisely chose science fiction and a backdoor attack on evolution using logic, philosophy and even poetry, which is obvious in much that he wrote.

---

6. Lewis, 2002, pp. 62-63. Italics in original.
7. Lewis, 2002, p. 63.
8. Burson and Walls, 1998, p. 31.

I encountered this in much of my reading of Lewis. His writings are much more understandable when reading with this insight. Lewis realized that the world painted by evolution was very different than the world painted by theism, and he wanted to spell out the differences in his books in an effort to counter the Darwinian worldview.

Much later in his life, after extensive reading and study, he had a very different attitude about science, writing that the Renaissance was not about the return to Greek science and empiricism as commonly believed, but rather was in fact more like a return to the

> golden age of magic and occultism. Modern writers who talk of 'medieval superstitions' 'surviving' amidst the growth of the 'scientific spirit' are wide of the mark. Magic and 'science' are twins *et pour cause,* for the magician and the scientist both stand together, and in contrast to the Christian.[9]

His longtime close associate, Walter Hooper, wrote that to appreciate Lewis' impact on Great Britain requires understanding the context in which he (Lewis) wrote, adding that "for centuries the heart of traditional Christianity had lain in the historic Incarnation, the belief that God became Man, but by the time Lewis was converted to Christianity in 1931, Agnosticism was everywhere."[10] So was the belief that nothing existed "but atoms and evolution," which the scientists claimed, or at least implied, explained everything.[11] The result of his reading and education was his belief that the

> reliability of biblical scholarship had been punctured a good many years before by a number of things, amongst which were those 'prophets of enlightenment': Charles Darwin, Karl Marx, Friedrich Nietzsche, Sig-

---

9. Lewis, C. S. 2004. *The Collected Letters of C. S. Lewis, Volume 2: Books, Broadcasts, and War, 1931-1949.* San Francisco, CA: HarperSanFrancisco, p. 68.

10. Hooper, 1996, p. 23.

11. Lewis, 2002, p. 167.

mund Freud and Émile Durkheim—all of whose works Lewis ...probably [was]... introduced to [in college].[12]

Markos wrote that, until Lewis' 1931 conversion from atheism, he

> accepted without question the modernist paradigm. But after that life-altering event that transformed Lewis in heart, mind, and soul, he began slowly to question and doubt the evolutionary presuppositions upon which all of his previous knowledge had rested. Indeed, again and again in his major apologetical works, Lewis confronts his readers with a series of human phenomena that could not have evolved—that is to say, that could not have arisen from natural, material causes alone. At present, there are many Christian scholars and writers who have uncovered flaws in Darwinian evolution.[13]

Lewis had to deal with the question, when becoming a Christian "must you be a creationist? What about evolution?"[14] Another problem was that Freud was "all the rage" when Lewis was teaching, and Freud regarded all religion as a human creation and a mere illusion. Freudian psychology has now been replaced by other icons. Freud did acknowledge that if

> you were to expel religion from our European civilization you can only do it through another system of doctrines, and from the outset this would take over the psychological characteristics of religion, the same sanctity, rigidity, and intolerance, the same prohibition of thought in self-defense.[15]

---

12. Hooper, 1996, p. 23.
13. Markos, 2003, p. 40.
14. Pearce, 2013, p. xx.
15. Freud, Sigmund. 1961. *The Future of an Illusion*. Translated from the German by James Strachey. New York: Norton, p. 66.

We do not know if Lewis read Freud's book, or even part of it, but he did have the same idea, that science would replace religion and, to some degree, that science had developed the "same sanctity, rigidity, and intolerance, the same prohibition of thought in self-defense." This is a topic that we will visit often in this book because it frequently came up in Lewis' writings.

Ironically, the evolutionists attributed the evidence that Lewis gave for God to the results of evolution. For example, atheist Beversluis attributes the source of "natural desires" to evolution, writing that natural desires "are biological and instinctive—evolutionary adaptations that trigger appropriate responses to external stimuli and whose satisfaction is necessary for the survival of individual organisms and of the species of which they are members."[16]

Lewis' Opposition to Eugenics

One fact that "deeply disturbed Lewis" was the "controlling potentials of science" noting "science brought vivisection, eugenics, the atomic bomb, social science planning instead of moral planning, and reminded [us] of our embarrassing relationship with the primates. Scientific Morality? Scientific discovery made World War II worse than the War to End all Wars."[17] His concern was "Scientists ... argued for eugenics, suggested we all descended from the apes... all at the expense and to the detriment of religion and tradition."[18] In short, Lewis recognized that when evolutionism became

> entangled with European capitalism and overseas expansion, this momentous turn in Western thought began to assume darker forms. The appearance of Darwin's *Origin of Species* (1859) gave additional impetus to this process by accelerating the tendency, already well under way prior

---

16. Beversluis, 2007, p. 45.
17. Miller, 2003, pp. 11-12.
18. Miller, 2003, p. 18.

to Darwin, to think of differences within the human species in terms of a ladder of ascent from the 'primitive' to the most "civilized." As it filtered back into social thought, Darwinian theory seemed to provide a biological rationale not only for an unregulated capitalism that encouraged "survival of the fittest" at home but also for the political, economic, and social domination of "underdeveloped" peoples abroad.[19]

His concerns about Darwinism were expressed in his novels, such as *Out of the Silent Planet* where Lewis wrote that many scientists in his day and their followers' agenda included "sterilization of the unfit, liquidation of backward races (we don't want any dead weights), selective breeding… biochemical conditioning."[20] Furthermore, Lewis observed if the survival-of-the-fittest idea is true, "in the process of being fittest for survival, our species has to be stripped of all those things for which we value—of pity, of happiness, and of freedom."[21]

As Schwartz correctly observed, Lewis no doubt saw "a seemingly benign program for improving the human stock [which later evolved] into the genocidal nightmare of the Third Reich."[22] Lewis, as a creationist, accepted a worldview radically opposed to eugenics, namely that all men were descendants of Adam and Eve, thus falsifying the belief that "differences within the human species [existed] in terms of a ladder of ascent from the 'primitive' to the most 'civilized.'"[23] Lewis' belief in the Biblical Adam was made clear in his writings. One of many examples was from his Preface to *Paradise Lost* where he wrote:

> Milton himself gives us a glimpse of our relations to Adam as they would have been if Adam had never fallen. He would still have been alive in

---

19. Schwartz, 2009, p. 29.
20. *Out of the Silent Planet*, p. 40. Quoted in Schwartz, 2009, p. 29.
21. Hooper, Walter. 1996. p. 207.
22. Schwartz, 2009, p. 101.
23. Schwartz, 2009, p. 29.

Paradise, and to that 'capital seat' all generations 'from all the ends of the Earth' would have come periodically to do homage.[24]

Milton then speculates what Adam would have been like if he had never sinned, as summarized by Lewis

> Adam was, from the first, a man in knowledge as well as in stature. He alone of all men "had been in Eden, in the garden of God, he had walked up and down in the midst of the stones of fire." He was endowed, says Athanasius, with "a vision of God so far reaching that he could contemplate the eternity of the Divine Essence and the coming operations of his word." He was "a heavenly being" accustomed to converse with God "face to face." His mental powers, says St. Augustine, "Surpass those of the most brilliant philosopher as much as the speed of a bird surpasses that of a tortoise."[25]

Lewis comments: "If such a being [as Milton described above] had existed—and we must assume that he did before we can read the poem—then Professor Raleigh and, still more, myself, on being presented to him would have had a rude shock."[26]

Professor Schwartz noted that when the novel *Out of the Silent Planet* was written, we cannot know for certain that Lewis was fully aware of the Nazi "racial hygiene" programs based on eugenics, but we know he was very aware of the eugenics movement, not only the one in Great Britain, but also in America and elsewhere. He was also cognizant of the scriptural opposition to eugenics.[27] Lewis read the Catholic, G. K. Chesterton's book, *Eugenics and Other Evils* (1923),

---

24. Lewis, C. S. 1942. *A Preface to Paradise Lost*. New York: Oxford University Press, p. 118.

25. Lewis, 1942, p. 117. Also quoted in Wilson, A. N. 1990. *C. S. Lewis: A Biography*. New York: Norton, p. 210.

26. Lewis, 1942, p. 117.

27. Schwartz, 2009, p. 30.

"considered the first sustained critique of the movement." Lewis also referred several times in his diary to the then popular topic of eugenics, and mentioned he was reading Chesterton's *Eugenics* book.[28] Jews in Nazi Germany were often called diseased or vermin. Lewis wrote that Jews "were objects… killed not for ill desert but because, on [Hitler's and the Nazi's evolutionary inferiority] theories, they were a disease in society."[29] Lewis also was concerned about the problem of inequality, especially "in the brutal form of Nazi ideology."[30]

Lewis' opposition to eugenics was also reflected in the outset of his Space Trilogy when Mr. Devine and the scientist Weston were preparing to sacrifice the "idiot boy" Harry "who in a civilized community would be automatically handed over to the state laboratory for experimental purposes" as the Nazis did.[31] Weston's "ruthless evolutionary ethics makes him equally prepared to exploit or exterminate the aliens."[32]

Weston is "impelled by a seemingly impersonal ideal of human progress and regards his venture into space as a necessary step in the development of the [human] species… that a 'scientific' hope of defeating death is a real rival to Christianity." This is only one example of the "transposition of the principal locus of Being from a transient God to an immanent power that realizes itself in the dynamic development of Man… the new paradigm of 'Evolution' or 'Development' or 'Emergence,'"[33] all terms Lewis used to describe Darwinism, as per Ruse's

---

28. Schwartz, 2009, pp. 168-169.
29. Lewis, C. F. 1958. *God in the Dock: Essays on Theology and Ethics*. Grand Rapids, MI: Wm. B. Eerdmans. p. 313
30. Lewis, C. S. 1986. *Present Concerns*. New York: HarperCollins. p. 8.
31. Schwartz, 2009, p. 187.
32. Schwartz, 2009, p. 27.
33. Schwartz, 2009, p. 29.

definition documented in Chapter One of this book.

Yet another example is found in the opening volume of the Space Trilogy set. The evil kidnappers of Ransom "are associated with the popular 'materialistic' view of 'orthodox Darwinism'... the infamous 'struggle for existence' especially as it appears in H. G. Wells' dramatic portrayal of this conflict on an interplanetary scale in the *War of the Worlds* and elsewhere."[34]

A major difficulty atheist philosopher S. T. Joshi had with Lewis was that "Lewis had no problem with evolution when he maintained his atheist views. The problem came in September of 1931 when Lewis converted to Christianity."[35]

---

34. Schwartz, 2009, p. 10.

35. Quoted from *The Official Website of C. S. Lewis*. www.cslewis.com/us/about-cs-lewis.

The Founder of Eugenics. Darwin's Cousin Francis Galton. From the frontispiece of Pearson, Karl. 1914. *The life, letters and labours of Francis Galton.* Cambridge: Cambridge University Press. Lewis aggressively opposed Eugenics.

7

Return to Christianity and Opposition to Darwinism

AFTER MUCH self-soul-searching and intensive reading, Lewis crossed the great Rubicon and returned to his Christian faith at the age of 33. Lewis' conversion was so important to him and his readers that, when Lewis wrote his autobiography, it was "written partly in answer to requests that I would tell how I passed from Atheism to Christianity."[1] He added that the focus in his autobiography was to tell "the story of my conversion and is not a general autobiography, still less 'confessions.'"[2] When he converted from atheism to theism in his college room in 1929, he finally "admitted that God was God," and described himself as "the most dejected and reluctant convert in all England."[3]

What forced him to convert were the facts uncovered by his extensive reading and research, bolstered by reason. At that moment he

---

1. Lewis, 2002, p. vii.
2. Lewis, 2002, p. vii.
3. Wilson, 1990, p. 110.

recognized there were "two and two only supreme and luminously self-evident beings, myself and my Creator." Once he accepted God as his Creator, not evolution, the next step he took, two years later, was to accept Christianity.[4] No 'Damascus Road experience' occurred to him, but rather it required a lot of study because Lewis had to deal with the many reasons why he became an atheist. Specifically, one issue was the beliefs of what he called the many pagan religions around him.[5] Factors he noted that were important in this step included reading, study, and the input from his academic friends, specifically Professor J.R.R. Tolkien and Hugo Dyson.[6]

One other critical factor in his conversion was his study of apologetics that enabled him to reject atheism. This study included the writings of George MacDonald who used stories to convey Christian apologetics, just as Lewis would become famous for later. Lewis soon became the "leading popular Christian apologist of the 20th century."[7] Although the "most widely read religious spokesman of our time … his main occupation was scholarship and university teaching."[8]

As he studied the evidence and learned about the many problems with evolutionary theory, Lewis also increasingly rejected Darwinism. In a letter to Douglas Bush dated March 28th, 1941, Lewis recognized that "Darwin, …made an upheaval" in science and society greater than that of even Copernicus and Freud. Lewis was, for this reason, very concerned about Darwinism's effects on people and society.[9]

Lewis openly rejected the orthodox view of evolution, namely that

---

4. Wilson, 1990, p. 110.
5. Burson and Walls, 1998, p. 30.
6. Duriez, 2013, pp. 120-121.
7. Tolson, 2005, p. 48.
8. Bredvold, 1968, p. 1.
9. Lewis, 2004b, p. 476.

natural forces ultimately account for the creation of everything in the universe. Lewis wrote that his "Deepening distrust and final abandonment of [evolutionary cosmology] long preceded my conversion to Christianity. Long before I believed Theology to be true I had already decided that the popular scientific picture [evolutionary cosmology] ... was false."[10] As will be discussed in detail, it was in 1957 when he made his only public attack against the theory. In a poem titled *Evolutionary Hymn*, "he mocked evolutionary pretensions to be a religion leading us up the future's endless stair to who knows where."[11] He also mocked evolution, or in Shaw's words "derided Evolution" by writing "Evolutionary biology is 'the science of the everlasting transmutation of the Holy Ghost in the world.'"[12]

Professor Louis Markos, a University of Michigan Ph.D., is one of the leading Lewis scholars today. He is the author of four books on Lewis, plus a 12-hour lecture audio series with *The Teaching Company* titled *The Life and Writings of C. S. Lewis*. Markos wrote that if Lewis "were alive today, he would be an ID (intelligent design) ...[supporter and] would have seen the flaws in Darwin and probably taken up the ID cause."[13]

The books[14] that Markos published include *Restoring Beauty: The Good, the True, and the Beautiful in the Writings of C. S. Lewis*[15]; *Lewis*

---

10. Lewis, 1962, p. 162. In his paper "Is Theology Poetry?"
11. Balfour, Arthur James, and (editor) Michael W. Perry. 2000. *Theism and Humanism: The Book that Influenced C. S. Lewis*. Seattle, WA: Inkling Books, pp. 158-159.
12. Lewis, 1967a p. 300. The quote was from George Bernard Shaw in his book *Back to Methuselah*. P. xxxii.
13. Markos, Louis. 2012. E-Mail to Jerry Bergman, dated September 12.
14. https://www.goodreads.com/author/list/45527.Louis_A_Markos.
15. Markos, Louis A. 2010. *Restoring Beauty: The Good, the True, and the Beautiful in*

Agonistes: How C. S. Lewis Can Train Us to Wrestle with the Modern and Postmodern World[16]; and *On the Shoulders of Hobbits: The Road to Virtue in Tolkien and Lewis.*[17]

According to Professor Markos, a major concern of Lewis was responding to those who attempt to explain the existence of all life "by the process of evolution."[18] Lewis concluded that these persons "fail to explain why and how that process would evolve in us a yearning for realities that lie outside that very process and that cannot be satisfied by it," such as the existence of an afterlife.[19] Furthermore, for his entire life "Lewis displayed a healthy skepticism of the many claims made in the name of science. He expressed this skepticism shortly before he became a Christian."[20] For example,

> while still an unbelieving undergraduate in 1922, he recorded in his diary a discussion with friends where they expressed their doubts about Freud. In 1925, he wrote his father about his gratitude toward philosophy for showing him "that the scientist and the materialist have not the last word." The next year he published his narrative poem *Dymer*, which offered a nightmarish vision of a totalitarian state that served "scientific food" and "[c]hose for eugenic reasons who should mate."[21]

Lewis started working on *Dymer* (the name of the lead character)

---

the *Writings of C.S. Lewis*. Downers Grove, IL: InterVarsity Press.

16. Markos, Louis A. 2003. *Lewis Agonistes: How C. S. Lewis Can Train Us to Wrestle with the Modern and Postmodern World*. Nashville, TN: Broadman & Holman.
17. Markos, Louis A. 2012. *On the Shoulders of Hobbits: The Road to Virtue in Tolkien and Lewis*. Chicago, IL: Moody Press.
18. Markos, 2003, p. 41.
19. Markos, 2003, p. 41.
20. West, 2012, p. 11.
21. West, 2012, p. 11.

when he was only 17. J. M. Dent Company published it nine years later in 1926 under the pseudonym Clive Hamilton, Lewis' first name followed by his mother's maiden name. Furthermore, in 1932, only a few months after he formally converted to Christianity, Lewis wrote to his brother about his concerns over the fact that the Rationalist Press Association was publishing

> cheap editions of scientific works they thought debunked religion. Lewis said their efforts reminded him of the remark of another writer "that a priest is a man who disseminates little lies in defense of great truth, and a scientist is a man who disseminates little truths in defense of a great lie [Darwinism]."[22]

Many academics recognized Lewis' antagonism against Darwinism. For example, Professor Harold Bloom of Yale, an agnostic Jew who opposed Evangelicals and personally knew Lewis, even attending his lectures at Cambridge, wrote that…

> Lewis' *Mere Christianity* is a perpetual best-seller among American Evangelical Christians. His attitude towards Evolution is a touch more sophisticated than theirs, but differs from Creationism only in degree, not in kind. Indeed, Intelligent Design is a kind of parody of Lewis' general view of a Christian cosmos.[23]

Bloom also edited a book on C. S. Lewis in which he stated that he read some two dozen of the 40 books and book chapters written by Lewis in print then. In view of this fact, his conclusions about Lewis are based on a detailed knowledge of him and his writings.[24] An analogy that Lewis used to support his non-evolutionary view in *Mere*

---

22. West, 2012, pp. 11-12.
23. Bloom, Harold. 2006. *C. S. Lewis (Bloom's Modern Critical Views)*. New York: Chelsea House Publications, p. 2.
24. Bloom, 2006, p. 2.

Christianity which illustrates his support for Intelligent Design is

> God made us: invented us as a man invents an engine. ...He Himself is the fuel our spirits were designed to burn, or the food our spirits were designed to feed on. ...God cannot give us happiness and peace apart from Himself, because it is not there.[25]

Some of Lewis' thoughts on evolution are not always easy to follow. For example, he wrote about his theological conversation with Rev. Dr. Frederick Walter Macran (1866-1947), a member of Parliament in the House of Commons, who he called 'Cranny'. Cranny did not believe that Jesus was who he (Jesus) said he was, and Lewis responded to Cranny noting "he thought evolution had first of all tried successful types, then settled down to the development of one type, MAN: in the same way we had first had successive religions and would now settle down to the development of one. I wonder if the mastodon talk the same way."[26]

Cranny evidently believed that religion had evolved, and personified Darwinian evolution, adding that all but one religion, like the mastodon, became extinct. In contrast, Lewis was obviously not impressed with evolution as was Cranny. It is unfortunate that more of their conversations were not included in Lewis' diary to more fully grasp both Lewis' and Cranny's views.

Lewis realized that the truth or falsity of Christianity is not dependent on other beliefs. For example, the "myth of dying and rising corn gods in pagan religions did not nullify the claim that Jesus Christ died and rose again" from the dead.[27] The charge that Christianity is discredited by its similarity to other religions was answered by the rejection of the evolutionary worldview and by acceptance of the creation

---

25. Lewis, C. S. 1980. *Mere Christianity*. New York: Simon & Schuster, p. 54.
26. Lewis, 1991, p. 22.
27. Burson and Walls, 1998, p. 30.

worldview. Instead of "seeing similar motifs as a strike against the faith, Lewis persuasively insisted this is precisely what we ought to expect if God is really the Creator and Sustainer of the entire world."[28]

---

28. Burson and Walls, 1998, p. 261.

Owen Barfield as a young man. He had a profound influence on C. S. Lewis especially through his book *The Silver Trumpet and Poetic Diction* (dedicated to Lewis).

8

Enter Owen Barfield

ARTHUR OWEN BARFIELD (1898-1997), who later became a well-known British philosopher and author, had an important influence on Lewis when he (Lewis) was wrestling with the problem of the origins of everything, including life, and the implications of one's theory of origins on their worldview. Owen Barfield had known Lewis from the time of Lewis' return to Oxford after World War I. When they first met in 1919, Lewis was an atheist who informed Barfield as a matter of fact, "I don't accept God!"[1] And Lewis "buttressed his atheism with the sort of scientism and chronological snobbery that he would later condemn in the most forthright terms."[2] Barfield was not only influential in converting Lewis to theism, but also solidified Lewis' rejection of Darwinism. When they met, Barfield "didn't believe in [Orthodox] evolution.... . Now Lewis, as you know, hated the idea of evolution."[3]

Barfield was, in turn, very influenced by Rudolph Steiner (1861-1925), an Austrian philosopher who attempted to find a synthesis between science and spirituality. Although previously Steiner had lit-

---

1. Duriez, 2013, p .88.
2. Peace, 2003, p. 13.
3. Poe and Poe, 2006, p. 104.

tle connection with Christianity, in 1899 he experienced what he described as a life-transforming encounter with Christ.[4] Steiner helped

> Barfield see that there was no need to accept the Darwinian, purely materialistic interpretation of the world. The crude Darwinian view of human consciousness, for example, was that it [human consciousness] somehow or other 'evolved' from a succession of increasingly intelligent apes, beginning with a creature who little thought beyond where his next banana was coming from, and culminating in the President of the Royal Society. But this was only a theory and not, on the face of it, a particularly probable one.[5]

Wilson added that

> Lewis was sufficiently committed to the life of the mind to see that if what Barfield was saying was true, it would profoundly affect everything. There cannot be a greater difference than that between someone who supposes that the human race (and with it all art, philosophy, science and virtue) is [in the end] a mere atomic accident in a blankly meaningless universe and those who believe that there is a plan, and behind it all a design.[6]

During the summer of 1925, the American Scopes Monkey Trial drew intense national publicity as national reporters flocked to Dayton, Tennessee to cover the big-name lawyers who had agreed to represent each side of one of the most famous court trials in American history. The press coverage of the "Monkey Trial" over teaching human evolution was enormous, and almost all were very critical of the belief that

---

4. McDermott, Robert A. 1995. "Rudolf Steiner and Anthroposophy," in Faivre and Needleman, *Modern Esoteric Spirituality*, New York: Crossroad Publishing, p. 292.
5. Wilson, 1990, p. 87.
6. Wilson, 1990, p. 87.

God created humans, and that they did not evolve. The front pages of major newspapers, including *The New York Times,* were dominated by the case for days. More than 200 newspaper reporters from all parts of the country and two from London, England were in Dayton.[7] Twenty-two telegraphers sent out 165,000 words per day covering the trial, over thousands of miles of telegraph wires hung for the purpose.

Reportedly, more words were transmitted to Britain about the Scopes Trial than for any previous American event. The most famous vituperative attacks came from journalist H. L. Mencken, whose syndicated columns for *The Baltimore Sun* drew vivid caricatures of what he referred to as the ignorant local populace. He referred to those who rejected Darwinism as morons and chastised the "degraded nonsense which country preachers are ramming and hammering into yokel skulls."[8] Ironically, the only evidence for human evolution at the time of the trial was a few Neanderthal skulls, Peking Man, Java Man, and the fraudulent Piltdown Man skulls.

According to the media, only ignoramuses rejected human evolution, yet one rejecter, C. S. Lewis, had just been elected a Fellow of Oxford's Magdalen College. He was previously a tutor in philosophy, but at Oxford took on a new academic field, namely English literature.[9] No wonder Lewis was shy for years about openly rejecting evolution. He, nonetheless, very tactfully but effectively, opposed Darwinism.

Lewis' Aggressive Opposition to Darwinism

Lewis also took on new challenges at Oxford, namely Darwinism. As

---

7. Larson, Edward J. 1997. *Summer for the Gods: The Scopes Trial and America's Continuing Debate over Science And Religion.* New York: Basic Books, p. 213.

8. Ruse, Michael. 2001. *The Evolution Wars: A Guide to the Debates.* New Brunswick, NJ: Rutgers University Press, p. 90.

9. Sayer, George. 1994. *Jack: A Life of C. S. Lewis.* Wheaton, IL: Crossway Books, pp. 176-184; Green and Hooper, 1994, pp. 79-85.

Professor Brazier in the extensive three-volume study of Lewis' theology wrote,

> Lewis' idealism contradicted, to a degree, the belief (even an article of faith?) amongst 'modern' philosophers of the closed-off cosmos, the positivistic [materialistic] realism that asserted an accident of evolution as the source of all, and not a creator God... Wrong, asserts Lewis; what is more, creation as the work of Christ... is intimately interconnected with eternity—it issues from and is sustained from eternity."[10]

Furthermore, Lewis believed that "God created everything out of nothing and sustains it... this may be obvious but it separates Lewis from many modern theologians and churchmen who watered down the faith."[11]

An example of the conflict Lewis faced is one of his doctoral students, Alastair Fowler (1930-), who wrote that he (Fowler) occasionally had a few disagreements with Lewis, then his doctoral supervisor. One "concerned Charles Darwin; Lewis saw the theory of natural selection as threatening religion" whereas Fowler did not. Fowler added that, from their debates on evolution, it was obvious that Lewis had very strong feelings against Darwinism, so strong that he [Fowler] "grew as red as Lewis himself. But he [Lewis] nimbly reined in, avoiding the threatened collision; he never lost his temper in debate [with him about Darwinism]."[12]

Fowler's education had been on the "science side, leading to a year in medicine at Glasgow University; I thought I knew quite a bit about

---

10. Brazier, Paul. 2012. *C. S. Lewis: Revelation, Conversion, and Apologetics (C. S. Lewis: Revelation and the Christ, Book 2)*. Eugene, OR: Pickwick Publications (Imprint of Wipf and Stock Publishers), p. 184.

11. Brazier, 2012, p. 8.

12. Poe and Poe, 2006, p.104.

genetics" Fowler explained.¹³ After explaining his evidence, probing Fowler's views on evolution, Lewis, according to Fowler, answered by citing

> an argument from Philip Gosse's ill-fated Omphalos. "You talk about fossils. How do you know God didn't put the fossils in the rocks?" Lewis would assume I had read enough Gosse to see the wit of using the Victorian's subtle compromise to test ... modern science. Anyhow, I was furious. How could he ignore the evidence of the geological record? Or was that a plant, too? Did God often lie to us? Or maybe he was trying out the old argument as one might casually heft an ancient but still serviceable mace.

In the book *Omphalos* (Greek for *navel*, the idea that Adam was created with the appearance of having been born of a woman with an umbilical cord) written by zoologist Philip Grosse, a Young-Earth Creationist,¹⁴ Grosse proposed the idea that *ex-nihilo* creation always resulted in *ex-nihilo*-created objects having the *appearance* of age or having a history.

This conclusion also applies to all miracles, including everything from the creation of wine from water to the creation of both the scriptural record and scientific data. This idea also has a long history and, in modern times, was advocated by many creationists and Biblical scholars from Philip Gosse to Henry Morris. This view is also referred to as a "coherently mature universe," "functional completeness," "prochronism," meaning outside of time, "created fully grown," or the "appearance of maturity."¹⁵

---

13. Fowler, 2003. p. 70.

14. Gardner, Martin. 1957. *Fads & Fallacies in the Name of Science.* New York: Dover Publications, p. 124.

15. Fowler, 2003. p. 70.

The problems inherent in dating an instantaneous supernatural creation are illustrated by the creation of the first man, Adam.[16] If created as a mature adult, Adam would appear to an onlooker to be about twenty years of age when he was, in fact, only less than one day old. This does not imply that the Creator is deceptive, but reflects the fact that the human body *had* to be created fully formed and functional in order to exist as a living organism.

If Adam's blood was not already circulating in his circulatory system when he was created, the few minutes required to prime his circulation system could cause major cell death or damage. Furthermore, all of Adam's organs—including his heart, lungs, kidneys, brain, etc.—must have been functioning simultaneously as a unit the instant he was created. In other words, Adam must have been created *ex-nihilo* as a fully mature young man. Likewise, when Eve was created, Adam would have known that she was only a day or so old, yet she would have appeared as a fully mature adult able to have children.[17] After all, God's first command to them was "Be fruitful, and multiply, …" (Genesis 1:28a). Fowler, admitting he was wrong and Lewis correct, later wrote, he [Fowler] was full of

> "liberal" assurance that there could be no conflict between religion and science, I dismissed Lewis' question as willful obscurantism. If he was determined to set religion against Darwin, surely he could have found a better argument [than God planted the fossils to deceive us.]

Fowler added

> Many years later, when I read Omphalos, I was ashamed to find that Gosse had anticipated exactly the objections I made to Lewis in my ig-

---

16. Bergman, Jerry. 2011. "The Case for the Mature Creation Hypothesis." *CRSQ* **48**(2):168-177, Fall.
17. Poythress, Vern S. 2006. *Redeeming Science: A God-Centered Approach.* Wheaton, IL: Crossway Books, p. 117.

norance. Gosse is sometimes misrepresented as arguing that fossils were inserted to test faith, whereas in fact he revered the fossil record as revealing, without deception, God's laws of biological development. To reconcile this with biblical chronology, Gosse speculated that fossils "may possibly belong to a prochronic development of the mighty plan of the life-history of the world." Lewis must have realized I didn't know Omphalos, and could have crushed my argument by pointing this out; but the "bully" was too kindly for that. After my outburst I was less in awe of Lewis; his opposition to Darwin came over as simplistic. More recently, I have begun to see that evolution is more complex than it seemed then.[18]

Fowler, ever the Darwinist, concluded, "I still think Lewis failed to enter the world of modern science, probably through not grasping its mathematical character. He had so little grasp of mathematics that he could never pass the elementary algebra in Responsions, the Oxford entrance exam."[19] Of course, math, while helpful in physics, is not required to understand the fossil record or the other arguments used to support evolution, such as the 'useless organ' or 'poor design' arguments.

It was this contact, and other similar experiences with people highly trained in science, that caused Lewis to realize that, due to his lacking a solid science background, it was difficult for him to debate evolution with someone like Fowler, who had an excellent science background, although Fowler admitted that Lewis did much better defending his anti-Darwin position than he had first assumed. Lewis, though, realized that he could not debate genetics or the details of the fossil record against highly trained scientists. He thus decided he would focus on the harm Darwinism did to society and especially the Church. He did an excellent job in this area, so ignoring the biology and paleontology was a wise move.

---

18. Fowler, 2003. p. 70-71.
19. Fowler, 2003. P. 71-72.

Other Evidence of Lewis' Anti-Darwinism

A Master of Arts thesis on Lewis and science concluded that Lewis "later accepted the biblical view of man as being created in the image of God and, though fallen from the original position of fellowship with God [man was] ... of such value that the Incarnation and the Crucifixion resulted."[20] The thesis concluded Lewis taught that, in contrast to Orthodox Darwinism, that

> naturalism, determinism, and empiricism all view man as a biological accident with no meaning and no unmeasurable qualities like soul or spirit. They presuppose that there is no God nor absolute truth or values. The study of man by sociologists, psychologists, and anthropologists has reduced him to a thing having little dignity or worth, having little or no individual responsibility for his choices and actions.[21]

Lewis' major concern was, in attempting to explain the origin of humans,

> scientists have "explained away" his value and meaning. This sort of explaining, or reductionism, is contemporary nihilism, according to Swiss psychiatrist Viktor Frankel, who blames the complaint of meaninglessness heard from his patients on those who say "time and again that something is nothing but something else."[22]

Along with reductionism, "Lewis also deplored the arrogance of some scientists and philosophers who insist .... only that which may be verified empirically is true and real."[23] If it cannot be verified by empirical methods, it is not, from a science perspective, real.

---

20. Crowell, 1971, p. 73.
21. Crowell, 1971, p. 74.
22. Frankl, Viktor E. 1969. "'Nothing But--' On Reductionism and Nihilism." *Encounter* 33:54, November.
23. Crowell, 1971, p. 74.

DePuw University President Erik J. Wielenberg wrote that it is clear from Lewis' own writings that he (Lewis) accepted the design argument for God. Lewis' case for Christianity and God contains two main components. The first

> consists of arguments for the claim that there is, in addition to the natural, physical universe that we perceive with our senses, some transcendent being, a *Higher Power that created the natural universe* and *is "more like a mind than it is like anything else we know."*[24]

On May 13, 1943, the well-known Christian author Dorothy Leigh Sayers (1893-1957), wrote to thank Lewis for the insight that she gleaned from reading his writings.[25] In her letter, she also complained about an atheist who was communicating with her about his concerns relating to the validity of Christianity. She then asked Lewis if there existed

> any up-to-date books about miracles. People have stopped arguing about them. Why? Has Physics sold the pass? …Please tell me what to do with this relic [this atheistic descendant, liberal leftover, and theological throwback] of the Darwinian age who is wasting my time, sapping my energies and destroying my soul.[26]

This communication was one of several factors that motivated Lewis to begin his lifelong apologetics career. This career helped him nurture his concerns about evolution that grew as he became more knowledgeable. As will be documented, Lewis had major doubts about Darwinism not long after he became a believer, according to a 1927

---

24. Wielenberg, 2008, p. 56. Emphasis added.
25. She is most well-known in theology for her 1941 book *The Mind of the Maker* (HarperOne; Reprint edition, 1987).
26. Quoted in Hooper, Walter (editor). 1996. *C. S. Lewis: A Companion & Guide*. London: HarperCollins Publishers, p. 343.

letter referenced below, and these concerns became stronger as he pondered and researched the issue.

As early as 1927, after quoting a nineteenth-century scientist, Lewis wrote in a letter to his father about evolution, dated March 30th, in which he stated that to accept Darwinism, "you need more *faith* in science than in ... theology."[27] Since then, he did discuss, or at least mentioned, evolution, but always as an outsider to the field of life science, never committing himself as a believer in Darwinism. For example, instead of stating, "Evolution has been proven by the fossil record," Lewis wrote Scientists and others "infer Evolution from fossils."[28] Scientists and others do infer evolution from the fossil record which is quite different than Lewis stating he believed "Evolution based on the clear evidence of the fossil record." Or even "I am convinced of the fact of Evolution based on the evidence of the fossil record" as those who argue that he was a theistic evolutionist claim.

A review of a book on Lewis and science, edited by John West, explored Lewis' views on Darwinism. In this book, West lays to rest the myth that Lewis was a gung-ho theistic evolutionist.[29] West acknowledges that at one time in his early life, Lewis may have

> accepted the plausibility of some kind of common descent. However, later in life he became more skeptical about any form of evolution, though ... Two of the most interesting findings by West are: 1) that Lewis was skeptical of Darwinism before he even converted to Christianity; and 2) that Lewis consistently rejected one major feature of Darwinian evolu-

---

27. Lewis, C. S. 2004. *The Collected Letters of C. S. Lewis, Volume 1: Family Letters, 1905-1931.* San Francisco, CA: HarperSanFrancisco, p. 680.

28. Lewis, C. S. 2001. *Miracles: A Preliminary Study.* San Francisco: HarperSanFrancisco, p. 21.

29. Review by Richard Weikart. "C. S. Lewis and Science." *Credo Magazine* (October 24, 2012). http://www.credomag.com/2012/10/24/c-s-lewis-and-science.

tion: its insistence on random, non-teleological processes. Even when he [appeared to have] accepted the possibility of common ancestry, he always rejected the notion that the process could have been random. West argues that Lewis' position is thus much closer to Intelligent Design than to Darwinism.[30]

This brief review is a good summary of Lewis' lifelong attempt to grapple with the entire Darwinian evolution issue. The following pages continue this review.

---

30. Review by Richard Weikart. "C. S. Lewis and Science." *Credo Magazine* (October 24, 2012). http://www.credomag.com/2012/10/24/c-s-lewis-and-science.

G. K. Chesterton at age 17 taken around 1908. Chesterton had a profound influence of C. S. Lewis. Lewis considered Chesterton's rebuttal to H. G. Wells Darwinism arguments a significant contribution in his own conversion away from evolution and into Christianity.

9

G.K. Chesterton's *The Everlasting Man*

BOOKS THAT INFLUENCED Lewis' opposition to Darwinism include *The Everlasting Man*, first published in 1925 and written by one of the most well-known nineteenth-century anti-evolutionists, Gilbert Keith (G. K.) Chesterton (1874-1936). Lewis also recommended this book to others on the issue of evolution.[1] Evolutionist Martin Gardner called this book an anti-evolution missive that was written partly to rebut Darwinism, and especially to respond to H. G. Wells' arguments for evolution as contained in his wildly successful, still-in-print book, *Outline of History*.[2]

Gardner added that Chesterton's *Everlasting Man* book argued that the enormous gap existing between humans and apes strongly argues against human evolution.[3] Genetic analysis has shown that this gap actually is a chasm involving close to a billion genetic DNA base-pair

---

1. See C. S. Lewis letter to Joseph Canfield dated February 28, 1955.
2. Gardner, 1957, p. 134.
3. Gardner, 1957, p. 134.

differences.⁴ Wells, in his *Outline,* argued for the evolutionary origin of humans, focusing on Darwin's conclusion that humans are quantitatively, but not qualitatively, higher than the animals.

Lewis viewed *The Everlasting Man* book as so central to his Christian walk that he credited it with moving him from his previous half-converted state to fully embracing conservative Christianity. He wrote to educator Rhonda Bodle on December 31, 1947, that "the very best popular defense of the full Christian position I know is G. K. Chesterton's *The Everlasting Man.*"⁵ As late as 1961, he still regarded this book as an excellent apologetic tool, adding that for "a good ("popular") defense of our position against modern waffle ..., I know nothing better than G. K. Chesterton's *The Everlasting Man.*"⁶

Lewis first became aware of Chesterton's writings while in the hospital recovering from "trench fever" when serving as a solder during World War I. This was when he

> first read a volume of Chesterton's essays. I had never heard of him and had no idea of what he stood for; nor can I quite understand why he made such an immediate conquest of me. It might have been expected that my pessimism, my atheism, and my hatred of sentiment would have made him to me the least congenial of all authors. It would almost seem that Providence, or some "second cause" ..., quite overrules our previous tastes when it decides to bring two minds together.⁷

Lewis added, when reading Chesterton reminded him of Scottish

---

4. Tomkins, Jeffrey P. and Jerry Bergman. 2012. "Is the human genome nearly identical to chimpanzee?—a reassessment of the literature." *Journal of Creation* 25(4):54–60.
5. Lewis, C. S. 2008, *Yours, Jack: Spiritual Direction from C. S. Lewis.* New York: HarperOne, p. 125.
6. Lewis, C. S. 1977. *The Joyful Christian.* New York: Macmillan, p. 103.
7. Lewis, 2002, pp. 182-183.

writer and scholar George MacDonald (1824-1905), who believed that "all imaginative meaning originates with the Christian creator God,"[8] Lewis "did not know what he was letting himself in for. A young man who wishes to remain a sound Atheist cannot be too careful of his reading."[9] In Lewis' words, when he read Chesterton's *Everlasting Man* for the first time, he

> saw the whole Christian outline of history set out in a form that seemed to me to make sense. ... I already thought Chesterton the most sensible man alive "apart from his Christianity." Now I veritably believe, I thought—I didn't of course *say*; words would have revealed the nonsense—that Christianity itself was very sensible.[10]

In another letter, dated December 14, 1950 to Oxford history graduate student Sheldon Vanauken, Lewis called *The Everlasting Man* "the best popular apologetic I know."[11] The book was also cited as number two in Lewis' list of the ten books that were the most important in shaping his Christian faith and philosophy of life.[12]

To document his case for human evolution, Darwin penned the two-volume, nearly 900-page book, *The Descent of Man*, that Lewis wrote "forcefully argued that unguided natural selection could produce man's mental and moral faculties perfectly well.... Lewis thought otherwise, and he was tutored in his doubts by a book from one of his

---

8. Duriez, 2013, p. 183.
9. Lewis, 2002, p. 183.
10. Lewis, 2002, p. 213.
11. Lewis, 2002, p. 154.
12. "What books did most to shape your vocational attitude and your philosophy of life?" *The Christian Century* (June 6, 1962). Referenced in https://www.christiancentury.org/article/2011-05/c-s-lewis-s-aeneid.

favorite authors, ... Chesterton's *The Everlasting Man*."[13] Chesterton's writings, especially Chesterton's history of Christianity, were also important in Lewis' struggle to accept the Christian God as real. In short, the effect on Lewis from reading the anti-evolution book "*The Everlasting Man* was staggering. Ever since discovering Chesterton, Lewis had continued to read his works."[14] As Pearson writes, "The opening chapter of *The Everlasting Man* begins with a discussion of evolution and the limits of its application to any understanding of human history" then, quoting Chesterton, he adds "Nobody can imagine how nothing could turn into something."[15]

What the Book Is About

The book is a deliberate rebuttal of H. G. Wells' *The Outline of History*. According to the evolutionary history outlined by Wells and others, mankind is simply another animal. Chesterton disputes Wells' portrayals of human life as a seamless evolutionary development from animal life, to primitive humans, then to modern humans. In Chapter Two, titled "Professors and Prehistoric Men,"

> Chesterton skewered the pretensions of anthropologists who spun detailed theories about the culture and capabilities of primitive man based on a few flints and bones, likely inspiring Lewis' discussion of "the idolatry of artifacts" in *The Problem of Pain*. But Chesterton also provides in his book a full-throttled argument as to why Darwinism cannot explain the higher capabilities of man. In Chesterton's words, "Man is not merely an evolution but rather a revolution" whose rational faculties far outstrip those seen in the other animals.[16]

---

13. West, 2012, p. 128.

14. Pearce, 2003, p. 29.

15. Pearce, 2003, P. 28.

16. West, 2012, p. 128.

Chesterton mentioned the topic of evolution over 25 times in his book. As noted, a major focus of the book was to document the chasm of the mind (brain) between men and animals. One of his main points was that evolution is impossible because "there is such a huge difference between men and animals—men speak, create works of art, laugh, wear clothes, feel guilt, form governments, worship God , and so on."[17] Research today has only reinforced the existence of this chasm in spite of enormous amounts of money and time spent by Darwinists attempting to bridge the chasm.

One conclusion of Lewis from reading Chesterton was that non-scientific "or unscientific effects follow every major scientific paradigm shift, as Lewis points out in a pair of lectures he gave to an audience of scientists at the Zoological Laboratory, Cambridge, in 1956."[18]

A good example is the claim of "Laws of Nature". This concept was taken from an analogy with jurisprudence by classically-trained English philosopher and statesman Francis Bacon. Of course, there exist no **laws** of nature that can be rigorously proven, only consistency to the point that confirmed exceptions are explainable by other factors. For example, gravity is a "law", but helium-filled balloons move upward in air (for the same reason that objects less dense than water do not sink in water).

Likewise, not only empirical science, but science-in-name-only (such as the theory of evolution) can and does greatly influence the culture. This was a major concern of Lewis, as reflected in many of his writings, including his novels.

---

17. Gardner, 1957, p. 134.
18. Ward, Michael. 2013. "Science and Religion in the Writings of C. S. Lewis." A lecture given at The Faraday Institute, St. Edmund's College, Cambridge on Tuesday, 29 May 2012. Printed in *Science & Christian Belief* 25(1):3-16.

Thomas H. Huxley standing next to a blackboard where he has just drawn the skull of what he believed was one of our evolutionary ancestors, a gorilla. His vigorous public support for Charles Darwin's evolutionary naturalism earned him the nickname "Darwin's bulldog."

10

Lewis Openly Rejects Naturalism

ESSENTIAL TO ORTHODOX Darwinism is *naturalism*, from the word *natural* meaning not supernatural. Lewis defines Naturalism as the belief that nothing exists except Nature, the material world, a view also called materialism. Lewis was a supernaturalist, one who believes that, besides nature, something above nature exists, such as God and angels, supernatural referring to above nature.[1] Of course, in the end, the "naturalist's explanation is evolution."[2] Actually, the diehard naturalist's explanation *for everything* is ultimately evolution.

For Lewis, our ability to reason is derived from the divine reason. He concluded that "it is overwhelmingly more probable that mind will be produced by a previously existing mind than by a process such as evolution."[3] In contrast, orthodox evolution teaches that natural is what "comes forth, or arrives, or goes on, of its own accord," postulating that molecules eventually became people purely by natural law,

---

1. Lewis. C. S. 1996. *Miracles*. New York: Simon & Schuster, pp. 5-6.
2. Purtill, 2004, p. 44.
3. Purtill, 2004, p. 44.

time, chance, and a great deal of luck.[4] Lewis' argument is the same used by Intelligent Design proponents today, namely: The only source of information is an informational giver, namely an intelligent designer.[5] As Professor Purtill correctly concluded, Lewis' 'mental proof' of God "was a form of the design argument....God can account for that, whereas 'naturalism' cannot."[6] Purtill asks correctly, "Why are there 'laws of nature' at all?" an observation that is another example that laws imply a lawgiver.[7]

Lewis had a philosophical-literature education, not a science education, and writes he finds it "almost impossible to believe that the scientists really believe what they seem to be saying," that ultimately life arose from 'random' events. But, he adds, "it is the glory of science to progress" forward.[8] Questioning evolution further, Lewis notes, although "We infer Evolution from fossils," nevertheless "unless human reasoning is valid, no science can be true." even if based on physical entities such as fossils.[9] Furthermore, Lewis adds that for this reason "Naturalism ... discredits our processes of reasoning."[10]

One comment Lewis made about the wonders of the natural world attributed to Nature, or Mother Nature, as we often do, is that it appears "as if she had been designed for that very role" that she plays in creation. Lewis adds, "To believe that Nature produced God, or even

---

4. Lewis, 1996c, pp. 7, 17.

5. Reddington, Kenneth G. 2015. *Following the Truth, Wherever It Leads: An Investigation of What Is Reality (and How It Affects Our Lives)*. Eugene, OR: Wipf and Stock.

6. Purtill, 2004, pp. 44-45.

7. Purtill, 2004, p. 45.

8. Lewis, 1996c, p. 20.

9. Lewis, 1996c, p. 21. Capitalization of the word 'Evolution' is in original.

10. Lewis, 1996c, p. 22.

the human mind, is, as we have seen, absurd."[11] As for what he means by anthropomorphic "nature", Lewis is simply referring to orthodox evolutionism – as so often modern evolutionists do even today.

Why Lewis Opposed Naturalism

When Lewis moved to an academic post at Cambridge University, his first lecture there argued that "the so-called divide between a dark and ignorant Middle Ages and an enlightened Renaissance is mostly fictitious, invented by modern thinkers eager to discredit medieval faith and thought" that brought the West to the post Christian era.[12] As Lewis argued in this lecture, "Christians and Pagans had much more in common with each other than either has with a post-Christian. The gap between those who worship different gods is not so wide as that between those who worship [a god] and those who do not."[13] It is this gap that we will explore in this chapter.

Two topics of special concern to Lewis, especially when he was older and better informed about the subject, were orthodox materialistic Darwinism, which he came to regard as "a Great Myth," by which in this case he meant a false story, and "naturalism,"[14] which is the core of evolution.[15] Lewis often referred to "naturalistic theories of evolu-

---

11. Lewis, 1996c, p. 49.
12. Lewis, C.S. 1969. "*De Descriptione Temporum*" in *Selected Literary Essays*. Cambridge, MA: Cambridge University Press, p. 5. Also reprinted in *They Asked for a Paper*, 1962.
13. Lewis, 1969, p. 5. Also reprinted in *They Asked for a Paper*, 1962.
14. Martindale, W., and J. Root. (editors). 1990. *The Quotable Lewis*. Wheaton, IL: Tyndale House, pp. 193-197.
15. Ayala, Francisco J. 2007. Darwin's greatest discovery: Design without designer. *Proceedings of the National Academy of Sciences* 104:8567-8573, May 15, p. 8567.

tionary development" by the term *developmentalism*.[16]

He defined this term as the "extension of the evolutionary idea far beyond the biological realm: in fact, its adaption as the key principal of reality" such as in naturalism.[17] Some prefer the term Darwinism to describe this concept which is the dominant view today among many leading secular evolutionists. As will be discussed, Lewis concluded that naturalism was self-refuting because it was contradicted by both fact and reason.[18] He had no problem with what is today called microevolution, or, more accurately, variation within the Genesis kinds, an observation for which Lewis did "not think a Christian need have any quarrel."[19]

If, as Darwinism teaches, reasoning is behavior that "evolved entirely as an aid to" practice survival skills, then "No more theology, no more ontology, no more metaphysics… But then, equally no more Naturalism."[20] Lewis concludes that the theist is freed from evolutionary limitations and "is not committed to the view that reason is a comparatively recent [evolutionary] development molded by a process of selection which can select only the biologically useful. For him, reason—the reason of God… the human mind in the act of knowing is illuminated by the Divine reason."[21]

---

16. Lewis, 2007, p. 1584.
17. Lewis C. S. 1986. *Present Concerns*. New York: HarperCollins. p. 76.
18. Reppert, Victor. 2003. *C. S. Lewis' Dangerous Idea*. Downers Grove, IL: InterVarsity Press.
19. Lewis C. S. 1986. *Present Concerns*. New York: HarperCollins. p. 76.
20. Lewis, 1996c, p. 33.
21. Lewis, 1996c, p. 34.

Arthur J. Balfour. In 1962, Christian Century asked C. S. Lewis, to name the books that had most influenced his thought. Among those that Lewis listed was Arthur J. Balfour's 1915 book *Theism and Humanism* which challenged the foundation of Darwinism. Balfour was responsible for the 1917 Balfour Declaration committing Great Britain to the establishment of a Jewish homeland in Palestine. In short, Israel owes its existence to Balfour.

11

The Influence of Arthur J. Balfour

AN AUTHOR LEWIS listed in a 1962 *Christian Century* article[1] Lewis listed Arthur James Balfour, 1st Earl of Balfour (1848-1930) as an author that had a major impact on him. Balfour was the author of an anti-evolution book titled *Theism and Humanism: Being the Gifford Lectures Delivered at the University of Glasgow, 1914* by Arthur James Balfour, 1st Earl of Balfour (1848-1930). Balfour was one of Great Britain's most respected leaders, serving as prime minister from 1902 to 1905. Although most well-known today for the Balfour Declaration, which in 1948 established a homeland for the Jews in Palestine, he was a Christian apologist and outspoken anti-Darwinist whose ideas in the Gifford lectures permeate the first five chapters of Lewis' book *Miracles*.[2]

Theism and Humanism was Balfour's 1914 Gifford lecture present-

---

1. See footnote 250.
2. Balfour, Arthur James, and (editor) Michael W. Perry. 2000. *Theism and Humanism: The Book that Influenced C. S. Lewis*. Seattle, WA: Inkling Books, pp. 158-159.

ed at the University of Glasgow. The entire set of Gifford lectures seeks to show that the very best of humanity, whether aesthetical, ethical or epistemological, requires God for its foundation and as its main source of value.

In his second introductory lecture, Balfour discusses intelligent design and selection within the evolutionary process, a point he regards as bearing specifically on belief in the metaphysical in an age of science. He reasoned that the argument from design is plausible, *even in light of Darwinian evolutionary theory*. For Balfour, theism requires not only the design inference, but also a set of values, a knowledge of good, and a framework for reasoned thought. This, he argues, emanates from an examination of the human mind and the 'soul of man.'

After reviewing the theories of one of the leading evolutionists and social Darwinists, Herbert Spencer, Balfour wrote that a major problem with Spencer's evolutionary ideas in his book *Synthetic Philosophy* was "its central episode, the transition from the not-living to the living, was never explained by the author of the *Synthetic Philosophy*; and the lamentable gap must be filled in by each disciple [reader] according to his personal predilections."[3]

Balfour observed that "Spencer himself was, of course, no advocate of "design" after the manner of Paley; and I only mention his cosmic speculations because their unavowed optimism—the optimism that is always apt to lurk in the word 'evolution'—[is].... material peculiarly suitable for those who seek for marks of design in lifeless nature."[4]

He adds that "the great omission which mars the continuity of his world-story—[is] the omission ... of any account of the transition from the not-living to the living." In addition, "there are, besides this, two other omissions, one at the beginning of his narrative, and the oth-

---

3. Balfour, 2000, p. 28.
4. Balfour, 2000, p. 29.

er at the end, whose significance in relation to 'design' should receive a passing comment."[5]

Another problem for Darwinism that Balfour noted was that it ignores "the argument from 'design'" which was "the foundation on which those who use the argument have chiefly built" and have always sought for evidence of contrivances among life, such as the

> intricate adjustment of different parts of an organism to the interests of the whole; in the adaptation of that whole to its environment, they found the evidence they required. Arrangements which so irresistibly suggested purpose could not (they thought) be reasonably attributed to chance.

Furthermore, if we view

> organic adaptations and adjustments in themselves, scientific discovery has increased a thousand-fold our sense of their exquisite nicety and their amazing complexity. I take it as certain that, had no such theory as Natural Selection been devised, nothing would have persuaded mankind that the organic world came into being unguided by intelligence. Chance ...would never have been accepted as a solution. Agnosticism would have been scouted as stupidity.[6]

Balfour, who personally knew Charles Darwin, wrote that what changed the view he enumerated above, causing agnosticism to no longer be regarded as "stupidity" by the intellectual class, was Charles Darwin, who is

> regarded as the greatest among the founders of the doctrine of organic evolution; but there is nothing in the mere idea of organic evolution which is incongruous with design. On the contrary, it almost suggests guidance, it has all the appearance of a plan. Why, then, has Natural Se-

---

5. Balfour, 2000, p. 29.
6. Balfour, 2000, p. 30.

lection been supposed to shake teleology to its foundation?[7]

Balfour concluded, Natural "Selection does not make it harder to believe in design, it makes it easier to believe in accident; and ... design and accident are the two mutually exclusive alternatives between which the argument from design requires us to choose."[8] Thus, before Darwin "those who denied the existence of a Contriver [God] were hard put to ... explain the appearance of contrivance. Darwin ... provided an explanation" opening the door to evolution and atheism.[9] Furthermore, Balfour wondered, as evolution taught, could "the most complicated and purposeful organs gradually arise out of random variations, continuously weeded by an unthinking process of elimination?"[10]

If we "Assume the existence of living organisms, however simple, let them multiply enough and vary enough, let their variations be heritable, then, if sufficient time be granted, all the rest will follow. In these conditions, and out of this material, blind causation will adapt means to ends with a wealth of ingenuity which we not only cannot equal, but which we are barely beginning to comprehend."[11]

By this reasoning, Balfour opines, Darwin murdered the "Contriver," which is God. Balfour then concluded that the "theory of selection thus destroys much of the foundation on which, a hundred years ago, the argument from design was based. What does it leave untouched?"[12] He deals with this problem by observing that Natural

> Selection may modify these conditions, but it cannot start them ... it

---

7. Balfour, 2000, pp. 30-31.
8. Balfour, 2000, pp. 30-31.
9. Balfour, 2000, p. 31.
10. Balfour, 2000, p. 31.
11. Balfour, 2000, p. 31.
12. Balfour, 2000, p. 31.

may enable organic species to adapt ... But it cannot produce either the original environment or the original living matter. These must be due either to luck or to contrivance; and, if they be due to luck, the luck ... is great. ...We cannot measure the improbability of a fortuitous arrangement of molecules producing not merely living matter, but living matter of the right kind, living matter on which selection can act.[13]

Balfour added that, suppose we measure the odds against the accidental emergence of the required pre-life,

protoplasm, how are we to compare this probability with its assumed alternative—intelligent design? Here, I think, even [agnostic] Laplace's calculator would fail us; for he is only at home in a material world governed by mechanical and physical laws. He has no principles which would enable him to make exhaustive inferences about a world in which other elements are included.[14]

Thus, Balfour notes, the problem Darwin had, which still is a major problem today, is not the survival of the fittest, but rather the *arrival* of the fittest. Balfour writes that the progress achieved in either science or philosophy has not brought us closer to the solution of

the argument from design. ...Those who refuse to accept design do so because they think the world-story is at least as intelligible without it as with it. This opinion is very commonly associated with [the idea that] ... the laws of matter and energy are sufficient to explain, not only all that is, but all that has been or that will be. ...The choice, therefore, is not between two accounts of the universe, each of which may conceivably be sufficient. The mechanical account ... [therefore] doubly fails to provide a satisfactory substitute for design.[15]

---

13. Balfour, 2000, p. 31.
14. Balfour, 2000, p. 31.
15. Balfour, 2000, pp. 33-34.

Darwinism requires us to believe "that the extraordinary combination of material conditions required for organic life" is ultimately due to chance, however the fact is that

> material conditions are insufficient, and have somehow to be supplemented. We must assume ... an infinitely improbable accident, ... the case is even worse—for the laws by whose blind operation this infinitely improbable accident has been brought about are, by hypothesis, mechanical; and, though mechanical laws can account for rearrangements, they cannot account for creation; since .... consciousness is more than rearrangement; its causes must be more than mechanical.[16]

Balfour concluded "the common-sense "argument from design" ... carries us beyond mechanical materialism, it ... is inconsistent with naturalism [and] it is inconsistent with Agnosticism. ... the universe, or part of it, showed marks of intelligent purpose."[17] For "Balfour, a theism which is worth centering a religion around requires not only a design inference, but also a set of values, a knowledge of the good, and a framework for reasoned thought."[18] The reality that Lewis listed this book as one of the most important he had ever read indicates that he agreed with Balfour's basic intelligent design and anti-Darwin arguments reviewed above.

Balfour's Ideas Reflected in Lewis' Novels

In his 1945 novel, *That Hideous Strength*, a Florida Edwardian mansion called Belbury was acquired by "a group of corrupt scientists seeking to remake the human race by purging it of the traditional values of

---

16. Balfour, 2000, pp. 33-34.
17. Balfour, 2000, p. 34.
18. "Theism and Humanism." https://www.giffordlectures.org/lectures/theism-and-humanism.

freedom and dignity."[19] This, and their hostility to Christianity, was specifically the concern Lewis had with scientists, both in his day and in the future. The people at Belbury justified this goal by "the fact that it [the hostility] is occurring, and it ought to be increased."[20] The logical extension of this idea "means the loss of all traditional values: murder, cruelty, theft, violation of rights are 'justified' because they are happening."[21] As historian Richard Weikart explains:

> In *The Abolition of Man*, as well as in *That Hideous Strength*, C. S. Lewis lays bare the moral implications of the modern secular worldview that sees humanity as simply a cosmic accident. He also notes the hypocrisy of those modern thinkers who urge moral relativism on others, while maintaining their own alternate, but sometimes disguised, forms of morality.[22]

Professor Pelser writes that Lewis' non-fiction book *The Abolition of Man* was

> one of the most influential books of the twentieth century. In it, Lewis prophetically envisioned that the growing acceptance of moral subjectivism and the moral skepticism it breeds would lead to a pernicious moral and eugenic manipulation of mankind... If we are to forestall the abolition of man that Lewis envisioned, we must reject moral subjectivism.[23]

The title "the abolition of man" reflects Lewis' concern over, not

---

19. Duriez, 2013, p. 220.
20. Williams, Donald T. 2006. *Mere Humanity: G.K. Chesterton, C. S. Lewis, and J. R. R. Tolkien on the Human Condition*. Nashville, TN: B&H Books, p. 89.
21. Crowell, 1971, p. 65.
22. Richard Weikart. 2019. Whatever Happened to Human Rights?: Morality after C. S. Lewis' *Abolition of Man*. *Christian Research Journal*, December 17.
23. Pelser, Adam C. 2017. "The Abolition of Man Today." *CHRISTIAN RESEARCH JOURNAL* 40(2), April 24.

only evolutionary "eugenics and biotechnologies that violate the dignity of human beings," but also his concern that the science of evolution was in the end murdering God. Furthermore, Lewis first delivered the lectures that became the book *The Abolition of Man* in 1943, only

> two years before Germany surrendered to the Allied Powers in World War II. Had the Nazis been victorious, the second half of the twentieth century likely would have been marked by the global implementation of eugenics programs aimed at ridding the world once and for all of whatever classes of people were deemed "undesirable" or "unfit" by the ruling political and scientific powers. The Holocaust would have been just the beginning. And Germany wouldn't have been the only culprit. Hitler and the Nazis were not the only practitioners of eugenics in the twentieth century. In fact, the Nazis were inspired in their programs of forced sterilization and genocide by the legally sanctioned eugenics programs that were already being implemented across the United States well prior to WWII.[24]

And in 1945, the year that World War II ended, Lewis published *That Hideous Strength*, which was the

> third book in his science fiction trilogy. In that work Lewis depicted the dangerous consequences of embracing secular worldviews. His warning came at a time when Stalin and Hitler had committed horrific atrocities in the name of secular worldviews. Stalin, in the name of a Marxist worldview, slaughtered millions in his collectivization campaign and in the Great Purge. Marx, based on his atheistic position, had promoted environmental determinism, the view that human behavior is shaped by the environment. Marx, Lenin, and Stalin all believed that by altering the economy — specifically by eliminating private property — they could transform human nature, thus leading us into a society free from oppression, poverty, and strife.[25]

---

24. Pelser, 2017.
25. Weikart, 2019.

Furthermore, the Marxist worldview concluded that "objective morality and human rights are non-existent. Marxists believed that morality was a tool of bourgeois oppression. Consequently, they did not believe in objective human rights."[26]

In Lewis' novel of the future society, technocratic elites establish the National Institute of Co-ordinated Experiments, whose acronym is N.I.C.E. Their experiments are designed to control humans, and they are unconcerned about morality or human rights. One N.I.C.E. official stated, "If Science is really given a free hand it can now take over the human race and re-condition it: make man a really efficient animal."[27] The biological part of N.I.C.E.'s program included the Darwinian inspired eugenic

> "sterilization of the unfit, liquidation of backward races...[and] selective breeding." As shocking as these proposals may seem, leading thinkers in Britain, the US, and elsewhere were actively promoting such ideas at this time. Another way that N.I.C.E. officials hoped to get rid of those they deemed the riff-raff of humanity was by fomenting wars in such a way that the biologically inferior would perish. In addition to this program of biological elimination, N.I.C.E. would use environmental manipulation on the remaining humans. Everyone would be subject to psychological conditioning, which could include suggestion, or — for those not so pliable — threats or even torture. Persons engaging in bad behavior would not be punished, but rather reconditioned.[28]

Another example is a quote from *The Problem of Pain* that shows, generally speaking, Lewis had little respect for Enlightenment propaganda. Specifically, Lewis noted that it would be a mistake to believe

---

26. Weikart, 2019.
27. Lewis, C.S. 1974. *That Hideous Strength: A Modern Fairy-Tale for Grown-Ups.* New York: Scribner, p. 39.
28. Weikart, 2019.

that our ancestors were ignorant and therefore entertained pleasing illusions about nature which the progress of science has since dispelled. For centuries, during which all men believed, the nightmare size and emptiness of the universe was already known. You will read in some books that the men of the Middle Ages thought the Earth flat and the stars near, but that is a lie. Ptolemy had told them that the Earth was a mathematical point without size in relation to the distance of the fixed stars – a distance which one mediaeval popular text estimates as a hundred and seventeen million miles.[29]

Conclusions

Lewis listed Balfour's writings, which were openly anti-Darwin and in support of intelligent design, as among his top ten favorite books because he agreed with them and their ideas. This was no small compliment, given that Lewis was a voracious reader, probably devouring many thousands of books in his lifetime. The implications of Balfour's ideas were accurately expressed in several of Lewis' novels, including *The Abolition of Man* and *That Hideous Strength*.

---

29. Lewis, 1996d, p. 4.

Lewis was very concerned with the Darwinism ideas as advocated by H. G. Wells. Wells was one of the most popular writers when Lewis was alive. Shown above is on the cover of *Time Magazine* September 20, 1926.

12

The Darwinism that Lewis Rejected

THE DARWINIAN WORLDVIEW of many leading scientists dominates the scientific literature today. For example, one of the most prominent scientists today is Francisco Ayala. Professor Ayala, a disgruntled former Catholic Priest turned atheist, wrote that Darwin's greatest discovery was to prove that the design everywhere in nature does not require a designer because Darwin's theory of evolution dispensed with the need for an Intelligent Designer, i.e., God.[1] This conclusion Ayala articulated is a major reason why the vast majority of prominent scientists are atheists, or at least agnostics.

A survey by Larson and Witham found 93 percent of the members of America's most elite body of scientists, *The National Academy of Sciences*, are agnostics or atheists, and only seven percent believe in a personal God. As of 2021, this percent is close to the exact reverse for the American public as a whole.[2] The effects of Darwinism on theism,

---

1. Ayala, 2007, p. 8567.
2. Larson, Edward. J., and Larry Witham. 1998. Leading scientists still reject God. *Nature* 394:313, July 23.

as defined in this book, were effectively articulated by the leading evolutionary biologist today, Richard Dawkins. He explained, in a 1996 meeting that occurred in Clare College, that he had interviewed his good friend, Nobel Laureate

> Jim Watson, founding genius of the Human Genome Project, for a BBC television documentary ... I asked Watson whether he knew many religious scientists today. He replied: 'Virtually none. Occasionally I meet them, and I'm a bit embarrassed [laughs] because, you know, I can't believe anyone [today] accepts truth by revelation.'[3]

Darwinism explains design without the need for a designer; thus, if you accept Darwinism, no reason exists to believe in God. The major reason people give for belief in God is that creation requires an intelligent creator, namely God. And once the need for a creator is negated, evolution is the only option to explain the origin of creation.[4]

An example is a leading early Darwinist, Darwin's cousin Francis Galton. Galton, the founder of the eugenics movement, rejected theistic evolution after reading Darwin's 1859 book *The Origin of Species*, believing that "evolution was not divinely directed... man might just as easily evolve backward toward the apes as forward into the image of his once-fancied Creator. Man's true religious duty, therefore ... should be to [facilitate] the deliberate and systematic forward evolution of the human species" by eugenics.[5]

Evolution refers not only to the evolution of life from some one-celled first cell, but also the worldview used to explain the origin of the entire physical universe. According to this view, no need exists to invoke God because natural law, time and chance can, and did, create ev-

---

3. Dawkins, Richard. 2006. *The God Delusion*. Boston: Houghton Mifflin, p. 99.
4. Shermer, Michael. 2000. *How We Believe*. New York: Freeman, p. xiv.
5. Beavan, Colin. 2001. *Fingerprints: The Origins of Crime Detection and the Murder Case that Launched Forensic Science*. New York: Hyperion, p. 100.

erything physical that exists. Evolution, its adherents believe, explains everything existing in the universe: the stars, the galaxies, the solar system, the planets, the moons, and all life from amoebae to humans.

In a magazine titled *Destinies,* an editorial[6] also uses the term evolution to describe "the universe's evident evolution towards self-awareness... evolving ourselves and our artifacts to higher and higher levels of awareness to be fruitful and multiply, to populate the universe." Lewis opposed this view because it ignored the "suffering and despair of countless human beings... Lewis' was one of the first voices raised against 'interplanetary imperialism'; his voice remains one of the most eloquent." He speculated if we encounter intelligent beings somewhere else in the universe, will we "exploit and ruin them as contemporary Western civilization has done with so many ancient cultures?"[7] The existence of life on other planets was a possibility that Lewis did not discount, but attempted to deal with this possibility from a Christian point of view – just in case. However, so far it hasn't, but in the 1950s and 1960s, many people saw evidence of life on Mars, or some other planet, as a real possibility. No evidence of life has been found on any planet in our Solar System and, so far, outside of it in spite of spending decades and billions of dollars searching.[8]

One of the most succinct and widely disseminated definitions of evolution by natural selection was articulated by the late Cornell Professor Carl Sagan, who wrote both in his book *Cosmos* and in the script of his highly successful film series, also titled *Cosmos*, that "The cosmos is all that is or ever was or ever will be."[9] Richard Dawkins in *The*

---

6. Baen, James (editor). 1979. *Destinies*. Volume 1, no. 3, April-June. New York: Ace Books.
7. Purtill, 2004, p. 122.
8. Purtill, 2004, p. 122.
9. Sagan, Carl. 1980. *Cosmos*. New York: Random House, p. 4.

Blind Watchmaker defined evolution (specifically, natural selection) as

> the blind watchmaker, blind because it does not see ahead, does not plan consequences, has no purpose in view. Yet the living results of natural selection overwhelmingly impress us with the appearance of design as if by a master watchmaker, impress us with the illusion of design and planning."[10]

One of the world's leading evolutionists, the late Professor Theodosius Dobzhansky, defined evolution as a theory that comprises all of

> the stages of the development of the universe: the cosmic, biological and human or cultural developments. Attempts to restrict the concept of evolution to biology are gratuitous. Life is a product of the evolution of inorganic nature, and man is a product of the evolution of life.[11]

Sir Julian Huxley wrote soon after Darwin published his *Origin of Species* in 1859 that

> evolution was soon extended into other than biological fields. Inorganic subjects such as the life-histories of stars and the formation of the chemical elements on the one hand, and on the other hand subjects like linguistics, social anthropology, and comparative law and religion, began to be studied from an evolutionary angle, until today we are enabled to see *evolution as a universal and all-pervading process*.[12]

Lewis openly opposed this evolutionary view. In response to the

---

10. Dawkins, Richard. 1986. *The Blind Watchmaker*. New York: W.W. Norton, p. 21.
11. Dobzhansky, Theodosius. 1967. "Changing Man." *Science*, **155**(3761):409-415, January 27.
12. Huxley, Sir Julian. 1955. *Evolution and Genetics*, Chapter 8, pp. 256-289 in *What Is Science?* (Edited by James R. Newman), New York: Simon & Schuster, p. 272. Emphasis added.

evolutionary dilemma, Lewis wrote that the evolutionary view taught that our

> solar system was brought about by an accidental collision. Then the appearance of organic life on this planet was also an accident, and the whole evolution of Man was an accident too. If so, then all our present thoughts are mere accidents—the accidental by-product of the movement of atoms. And this holds for the thoughts of the materialists and astronomers as well as for anyone else's. But if *their* thoughts—i.e., of Materialism and Astronomy—are merely accidental by products, why should we believe them to be true? I see no reason for believing that one accident should be able to give me a correct account of all the other accidents.[13]

Conversely, the Myth, i.e., Darwinism, having "first turned what was a theory of change into a theory of improvement, then transformed into a cosmic theory. Not merely terrestrial organisms but *everything* in the cosmos is believed to be moving 'upwards and onwards.'"[14]

---

13. Lewis, 1970, pp. 52-53.
14. Lewis, 1967, p. 86.

NEW YORK TIMES BESTSELLER

The GOD Delusion

Richard Dawkins

Darwinism is the Doorway to Atheism, as argued by one of the most influential atheists today, Richard Dawkins. Most of Dawkin's books were written to defend evolution. *The God Delusion* pictured above is one exception. Book cover.

13

Lewis' Answer to Darwinism

ANOTHER TERM that must be discussed is creationism, which usually refers to the creation of all life kinds by God as briefly outlined in the Bible book of Genesis. A more restrictive definition includes belief in an Earth created from six-to-ten-thousand years ago and a global flood that was universal in its extent and effect.

I was unable to find any direct reference to these last two ideas, either pro or con, in Lewis' writings, or those of his associates who wrote about Lewis' beliefs and ideas. Thus, a reasonable judgment on his views in this area cannot be made. Lewis was likely not a strict YEC (Young-Earth Creationist), which was made possible for most people primarily through the work of Dr. Henry Morris and Princeton University graduate Dr. John Whitcomb. Together they wrote the seminal book, *The Genesis Flood*, first published in 1961. We have no evidence that Lewis ever read, or even owned, this book nor any other related book on the youth of Earth or the global extent of the Flood of Noah's day before his death in 1963. Nonetheless, Lewis often wrote as if the Creation was an instantaneous, *ex nihilo* event that occurred relatively recently.

In a chapter titled "Answers to Questions on Christianity". Lewis answered the following question:

> The Bible was written thousands of years ago for people in a lower state of mental development than today. Many portions seem preposterous in the light of modern knowledge. In view of this, should not the Bible be re-written with the object of discarding the fabulous and re-interpreting the remainder?[1]

Lewis' answer was as follows

> as to the people in a lower state of mental development. I am not so sure what lurks behind that. If it means that people ten thousand years ago didn't know a good many things that we know now, of course, I agree. But if it means that there has been any advance in intelligence in that time, I believe there is no evidence for any such thing.
>
> The Old Testament contains fabulous elements. The New Testament consists mostly of teaching, not of [p.58] narrative at all: but where it is narrative, it is, in my opinion, historical. As to the fabulous element in the Old Testament, I very much doubt if you would be wise to chuck it out. What you get is something coming gradually into focus. First you get, scattered through the heathen religions all over the world — but still quite vague and mythical — the idea of a god who is killed and broken and then comes to life again. No one knows where he is supposed to have lived and died; he's not historical. Then you get the Old Testament. Religious ideas get a bit more focused. Everything is now connected with a particular nation. And it comes still more into focus as it goes on. Jonah and the Whale, Noah and his Ark, are fabulous; but the Court history of King David is probably as reliable as the Court history of Louis XIV. Then, in the New Testament the ... dying god really appears — as a

---

1. *C.S. Lewis: Essay Collection and Other Short Pieces* Harper-Collins- chapter 4b Page 324. Also found on page 47 *God in the Dock*. Eerdmans 2014

historical Person, living in a definite place and time. If we could sort out all the fabulous elements in the earlier stages and separate them from the historical ones, I think we might lose an essential part of the whole process.

Looking up the word fabulous, The British Oxford English dictionary defines it as "extraordinary, especially extraordinarily large, amazingly good; wonderful." Unless evidence exists otherwise, I have to go with the standard definition. Some may define fabulous as false stories. But I see no good reason to assume this definition is warranted.[2]

Walter Hooper, Lewis' editor, stressed that Lewis was "a thoroughgoing Supernaturalist, believing in the Creation, the Fall, the Incarnation, the Resurrection, the Second Coming, and the Four Last Things (Death, Judgement, Heaven, Hell).[3] From this list, it seems Lewis would agree with the 40 percent of Americans, according to the 2019 Gallup poll, who accept these beliefs, including the view that God created the first humans within the last 10,000 years.

For example, in one of his *Chronicles of Narnia* books, in *The Magi-*

---

2. These questions were posed to C.S. Lewis on April 18, 1944 by workers at the Electric and Musical Industries Ltd., in Hayes, Middlesex. This company, now known as EMI, is best known for its recording studios at Abbey Road, London which was where the Beatles recorded their music. It's very interesting to see how Lewis handles their questions, most of the time keeping his answers quite short.

    Lewis may have been influenced in his growing critical view of Evolution by Dewar, Douglas 1949 *Is Evolution a Myth?: a debate between Douglas Dewar, L. Merson Davies and J.B.S. Haldane*. C.A. Watts/Paternoster Press London. One early criticism of the evolution proposed by C. S. Lewis in his essay "The Funeral of a Great Myth", was first published in 1945 and later included in the collection of essays: **Christian Reflections**. It is likely Lewis gained his critical view of evolution due to the influence of this pioneering UK creationist.

3. Lewis, 1967, p. vii.

cian's Nephew, Aslan created Narnia in a single day, not over millennia, or extended literary/geologic epochs. In his *Space Trilogy*, God created life, both sentient and mechanical, in *Perelandra*. God did this in a very short time, not over millions of years or throughout the long geological eras that are a central part of scientific orthodoxy today.

Intelligent Design theory did not exist as an organized movement when Lewis was alive, but Lewis did make many statements related to this worldview that help to determine if he would have accepted this view if he were alive today. Many persons, such as some Intelligent Design supporters, reject the evolution of all life from a common ancestor, but do not take a firm stand on a universal flood and a young Earth. As noted, several lines of evidence support the view that Lewis would have at least supported Intelligent Design today. For example, Lewis' support of Balfour's book being one of the most important books he had ever read, which indicates he agreed with Balfour's basic intelligent design arguments as reviewed in Chapter 11.

Intelligent Design focuses on evidence for intelligence in the natural world, such as the fact that the genomes existing in all life consist of information, and the only known source of information is intelligence. Due to several court rulings,[4] the movement is often referred to as Intelligent Design Creationism (IDC), which the media propagates as having been debunked, confusing the issue even more. To make matters even more difficult, Lewis *seemed* to waver a few times on the issue of origins, an issue that will be discussed in more detail.

A key focus of this treatise is the conclusion that "Lewis consistently rejected one major feature of Darwinian evolution," namely, naturalism and evolutionary progression by random, non-teleological processes. Thus, Lewis is much closer to Intelligent Design than to

---

4. Such as that in Pennsylvania Tammy Kitzmiller, et al. v. Dover Area School District, et al. (400 F. Supp. 2d 707, Docket No. 04cv2688).

Darwinism.⁵ In his lecture, "De Futilitate", presented at Magdalen College (Oxford University) during the Second World War, Lewis, when addressing a scientifically trained audience explained: "a much deeper and more radical futility" exists and this "cosmic futility is concealed from the masses by popular Evolutionism." Michael Ruse used the word "Darwinism" to describe the same worldview. Lewis explains:

> popular Evolutionism is something quite different from Evolution as biologists understand it. Biological Evolution is a theory about how organisms change. Some of these changes have made organisms, judged by human standards, 'better' – more flexible, stronger, more conscious. The majority of the changes have not done so. As J. B. S. Haldane says, in evolution *progress is the exception and degeneration is the rule*. Popular Evolutionism ignores this. For it, 'Evolution' simply means 'improvement'. And it is not confined to organisms, but applied also to moral qualities, institutions, arts, intelligence and the like. It is thus lodged in popular thought the conception that improvement is, somehow, a cosmic law: a conception to which the sciences give no support at all."⁶

A core belief of creationism is that what is observed in the natural world is not progression upwards from simple to complex life. In the talk to a scientific audience mentioned above, Lewis observed "there is no general tendency ... for organisms to improve. There is no evidence that mental and moral capacities of the human race have been increased since man became man. And there is certainly no tendency for the universe as a whole to move in any direction which we should call 'good,'" but rather it is it's movement in the direction towards "entropy, degradation, disorganization" which dominates.⁷ These conclusions would put Lewis firmly in the creationist/Intelligent Design camp and clearly

---

5. Weikart, 2012. http://www.credomag.com/2012/10/24/c-s-lewis-and-science.
6. Lewis, 1967, p. 58. From *De Futilitate*; emphasis added.
7. Lewis, 1967, p. 58. From *De Futilitate*.

argue against the view that he accepted biological evolution from fish to humans, or even the progression from apes to humans by means of improvements.[8]

Another of Lewis' statement that supports the creation/Intelligent Design camp is: "A great many people think ... Nature produced the mind. But on the assumption that Nature herself is mindless [which is the belief of science] this provides no explanation." Only mind can produce mind, and only intelligence can produce intelligence.[9]

Lewis' View of Science

Lewis stated that his acceptance of evolution was a major reason why he became an atheist. In his book, *The Problem of Pain*, Lewis wrote, "if anyone had asked me 'why do you not believe on God?' my reply would be" not because of the evil in the world, or the lack of proof for God, but rather because of the evidence of evolutionary cosmology.[10] By this he meant not that the facts or discoveries of science, such as the formula for water or the details of the Moon's surface geological construction, but of Darwinian beliefs about those facts in the terms that Michael Ruse explained (see chapter 1).

In short, as an atheist, Lewis believed, as taught then by orthodox scientists, that Earth "existed without life for millions of years and may exist for millions of years when life has left her."[11] He then repeats the evolutionary scenario that eventually led up to human consciousness. When he became a Christian, he rejected this Darwinian worldview and accepted the Christian Creation worldview.

Professor Schwartz, in a detailed study of Lewis' writings, provided further evidence that Lewis did not accept Darwinism as defined by

---

8. White, 1969, p. 191.
9. Lewis, 1967, p. 64. From *De Futilitate*.
10. Lewis, 1996d, p. 1.
11. Lewis, 1996d, p. 2.

Ruse, but did accept science fact, such as the observable physical traits of the Moon's surface. Schwartz concluded that Lewis had a good grasp of the

> seemingly impassable conflict between Christian tradition and the evolutionary or 'developmental' tendencies of modern thought. In his contemporaneous essays Lewis states repeatedly that his target is not the biological theory of evolution, which he regards as a "genuine scientific hypothesis," but the more deep-seated conceptual paradigm, well established by the time of Darwin's monumental *Origin of Species* (1859), which transferred the focal point from a transcendent God to the progressive development of Man" especially as applied to eugenics.[12]

Lewis was not concerned with so-called microevolution (or more accurately variation within a Genesis kind). His concern was the worldview called Darwinism, which "consigns other human beings to subhuman status, or summons up an 'evolutionary imperative' to legitimate... [eugenics] for the improvement of the species" as exemplified by the evolutionary application of eugenics which resulted in "Nazi aggression."[13]

Common Interpretations of Lewis' Views of Evolution

The three common interpretations of Lewis' views of evolution, as defined above, include:

1. He was a theistic evolutionist after he left atheism. This was the past view of Michael Peterson and others.

2. He was a theistic evolutionist after he left atheism, but became increasingly hostile to evolution in his later years.

3. He rejected theistic evolution for most of his life, although he

---

12. Schwartz, 2009, p. 6.

13. Schwartz, 2009, p. 6.

may have flirted with some evolution ideas, not necessarily because he agreed with them, but to make the point that Christianity could conceivably fit in with some forms of theistic evolution.

This book evaluates positions 2 and 3, and concludes that the overwhelming evidence supports view 3. Lewis is said to have never sidestepped tough issues but it took him years to be convinced of the need to take Darwinism head on.[14] It also took him years to openly confront his atheism. One feature worth commenting on in the poem Lewis authored before he became a Christian, *Loki Bound,* is its pessimism: "I was at this time living, like so many Atheists or Antitheists, in a whirl of contradictions." But "Loki Bound" was Lewis' most ambitious poetic endeavor in his still early development as a writer. Written in part before, and after, Lewis began his time with his free-thinking tutor W. T. Kirkpatrick in 1914, "Loki Bound" is a dramatic long poem maintaining that "God did not exist. I was also very angry with God for not existing. I was equally angry with Him for creating a world" with sin.[15] Furthermore, he writes that

> my rationalism was inevitably based on what I believed to be the findings of the sciences, and those findings, not being a scientist, I had to take on trust—in fact, on authority. Well, here was an opposite authority. If he had been a Christian I should have discounted his testimony, for I thought I had the Christians "placed" and disposed of forever. But I now learned that there were people, not traditionally orthodox, who nevertheless rejected the whole Materialist philosophy out of hand ... I [at the time] had no conception of the amount of nonsense written and printed in the world.[16]

---

14. Reid, 1990, p. 645.

15. Lewis, 2002, p. 110.

16. Lewis, 2002, p. 168.

In the poem, Lewis expressed "his contempt for the Christian view of the universe", i.e., creationism.[17] When Lewis accepted Intelligent Design, his tone was very different, writing that what "is behind the universe [namely God] is more like a mind than it is like anything else we know.... it [God] is conscious, and has purpose, and prefers one thing to another. And it [God] made the universe, partly for purposes ... in order to produce creatures like itself... to the extent of having minds."[18]

---

17. Latta, Corey. 2016. *C. S. Lewis and the Art of Writing: What the Essayist, Poet, Novelist, Literary Critic, Apologist, Memoirist, Theologian Teaches Us about the Life and Craft of Writing.* Eugene, OH: Wipf and Stock Publishers, p. 97.

18. Quoted in Beversluis, 2007, p. 74.

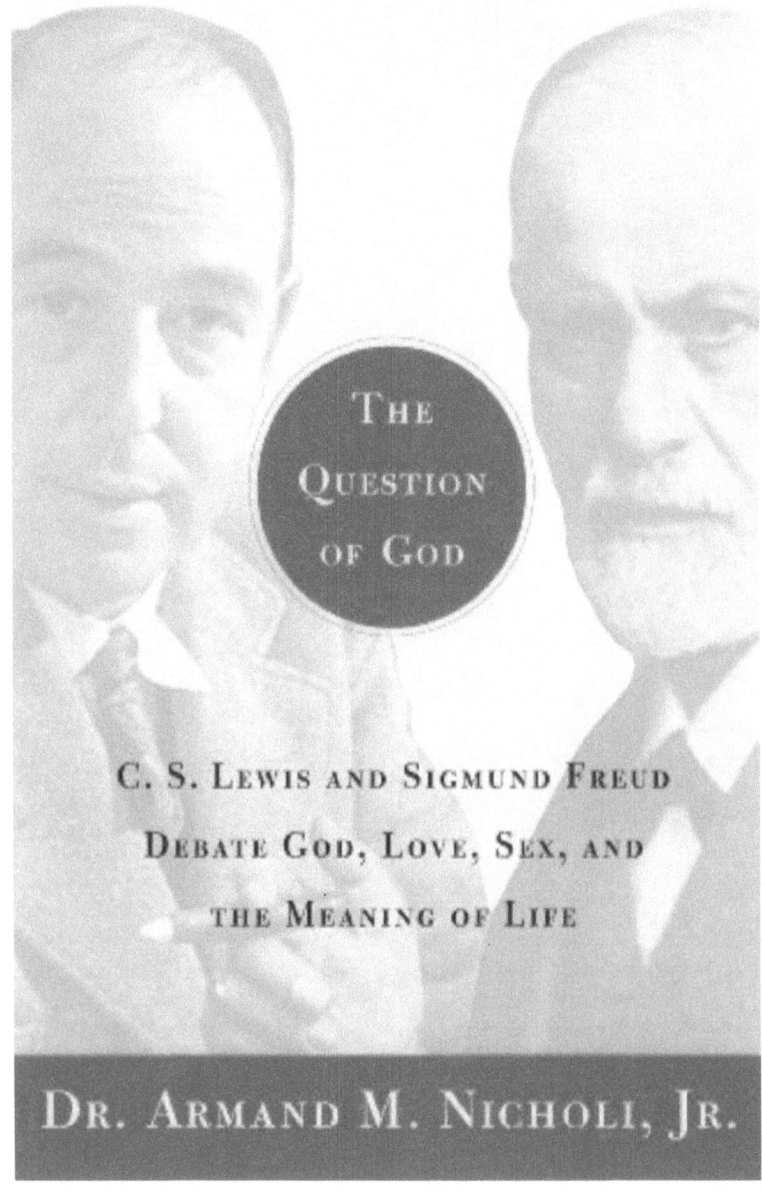

Harvard Professor Armand M. Nicholi book on C. S. Lewis' view of God, Darwinism, and the meaning of life. Book cover.

14

A Harvard Scholar's View of Lewis

FOR OVER 35 YEARS, the late Harvard University Professor of Psychiatry, Armand M. Nicholi (1927-2017), taught a course on Freud's person and ideas. He was also the editor and co-author of the classic *The Harvard Guide to Psychiatry*. The class eventually morphed into a course on both Freud and C. S. Lewis.[1] Nicholi's clinical work and research focused on the impact of absent parents on the emotional development of children and young adults. Dr. Nicholi later wrote a book, based on his Harvard course, contrasting and comparing the worldviews of these two intellectual giants. Actually, the book consists of *three* people: Freud and Lewis before, and Lewis after his conversion. As is obvious from his textbook, Nicholi is an expert on both men.

Professor Nicholi noted that both Freud and Lewis were reared in a religious environment, specifically Christian,[2] both became atheists as

---

1. Nicholi, Armand M. 2002. *The Question of God: C. S. Lewis and Sigmund Freud Debate God, Love, Sex, and the Meaning of Life*. New York: Free Press, p. 5.
2. Freud was born to Jewish-German parents, but went to a Catholic school and converted to Christianity as a youth.

adolescents, and both spent the rest of their lives proselytizing, Freud for atheism and Lewis for theism, specifically Christianity. Professor Nicholi documented in detail how important Intelligent Design was in Lewis' conversion to Christianity and how important Darwinism was in Freud's acceptance of atheism. In one chapter of his book, titled "*The Creator: Is There an Intelligence Beyond the Universe*," Nicholi writes that as

> an atheist, Lewis agreed with Freud that the universe is all that exists—simply an accident that just happened. But eventually Lewis wondered whether its incredible vastness, its precision and order, and its enormous complexity reflected some kind of Intelligence. Is there Someone beyond the universe who created it? Freud answers this "most important question" with a resounding "No!" The very idea of "an idealized Superman" in the sky—to use Freud's phrase—is "so patently infantile and so foreign to reality, that ... it is painful to think that the great majority of mortals will never rise above this view of life."[3]

Freud predicted that as the average common person becomes better "educated, they would 'turn away' from 'the fairy tales of religion.'" God, Freud argues, is an exalted father whom we invented in order to protect ourselves from the unfortunate vicissitudes of life. He reminds us that 'the world is no nursery' and strongly advises us to face the harsh reality that "we are alone in the universe."[4] In contrast, after his worldview changed, Lewis asserted in support of the Intelligent Design worldview

> that the universe is filled with "signposts" like the "starry heavens above and the moral law within"—Immanuel Kant's phrase—all pointing with unmistakable clarity to that Intelligence. Lewis advises us to open our eyes, to look around, and understand what we see. In short, Lewis shouts,

---

3. Nicholi, 2002, p. 36.
4. Nicholi, 2002, p. 36.

"Wake up!"[5]

One fact, Nicholi notes, that deeply impressed Lewis, was his observation that "our physical universe ... is extremely complex ... it comprises atoms, electrons, etc." and "the universe is not just the sum of its physical parts" but much more.[6] This reasoning is very much like that espoused by the modern Intelligent Design movement. Conversely, Freud believed that evolutionary science had shown that God is "so improbable, so incompatible with everything we had laboriously discovered about the reality of the world" by science.[7] As Nicholi pictures Freud, his lifelong tragic unhappiness made him rigid and unyielding and, as a result, he closed his mind to God and religion as well. Lewis' surrender and ultimate acceptance of God opened his mind to many possibilities as he repeatedly expressed in his writing.

Professor Nicholi writes that, in the 19th century, people turned to the discoveries of Darwin for "what they considered the irreconcilable conflict between science and faith."[8] Furthermore, "Freud's argument reflects the thinking of ... Darwin" which is where he [Freud] got the idea that men originally "lived in hordes, each member under the domination of a single powerful, violent and jealous male."[9]

Some of Freud's more questionable ideas, such as those contained in his book *Totem and Taboo: Resemblances Between the Mental Lives of Savages and Neurotics* "rested on a conjecture by Darwin that Primitive prehistoric people lived" under certain conditions that we humans inherited and which partly explains our evil behavior today. This work of Freud has now been fully discredited by anthropologists, specifically by

---

5. Nicholi, 2002, pp. 36-37.
6. Nicholi, 2002, p. 54.
7. Nicholi, 2002, p. 55.
8. Nicholi, 2002, p. 3.
9. Nicholi, 2002, pp. 42, 69, 71.

cultural anthropologist Alfred L. Kroeber.[10]

In contrast to Freud, Lewis did not resort to so-called prehistoric man or cultures, but "the plan of creation" that was designed for our happiness which was derailed by our first parents, Adam and Eve, whose sin was "simply and solely disobedience—doing what you have been told not to do."[11] Thus, the plan still exists. We just have to understand God's plan for us and live according to it. To help us find, and live by that plan, was the main goal of C. S. Lewis' writings. Conversely, Freud wondered how the "widespread belief in a Supreme Being 'obtained its immense power which overwhelms reason and science."[12]

Freud was very influenced by German physiologist Professor Ernst Wilhelm von Brücke (1819-1892), a man Freud claimed "carried more weight with me than anyone else in my whole life." He was "one of a group of physiologists who attempted to found a science of biology on thoroughly materialistic grounds", i.e., Darwinism.[13] Freud began studying under Brücke in 1877 and continued until 1883. Brücke was Professor of Physiology at the University of Vienna for over 41 years. It was from Brücke that Freud openly and aggressively advocated an atheistic philosophy of life [referring to this view] as the scientific Weltanschauung [worldview]."[14]

In contrast, Lewis did not believe biology, or even science, could "answer all questions, [thus] cannot be the source of all knowledge… the job of science … is to experiment and observe and report how

---

10. Kroeber, Alfred L. 1920. "Totem and Taboo: An Ethnologic Psychoanalysis". *American Anthropologist* **22**(1):48–55, January-March.
11. Nicholi, 2002, pp. 71, 122.
12. Nicholi, 2002, p. 103.
13. Nicholi, 2002, p. 20.
14. Nicholi, 2002, p. 2.

things behave or react."[15] Furthermore, Lewis believed "whether or not intelligence exists beyond the universe can never be answered by the scientific method."[16] Freud thought that it could be answered, and the answer was in the negative, i.e., there is no intelligence beyond the material universe. In short, the Darwinian secular worldview was central to the lives of both men – one accepting it and becoming an atheist, the other rejecting it and becoming one of the most popular Christian apologists of the last century.

---

15. Nicholi, 2002, p. 58.

16. Nicholi, 2002, p. 58.

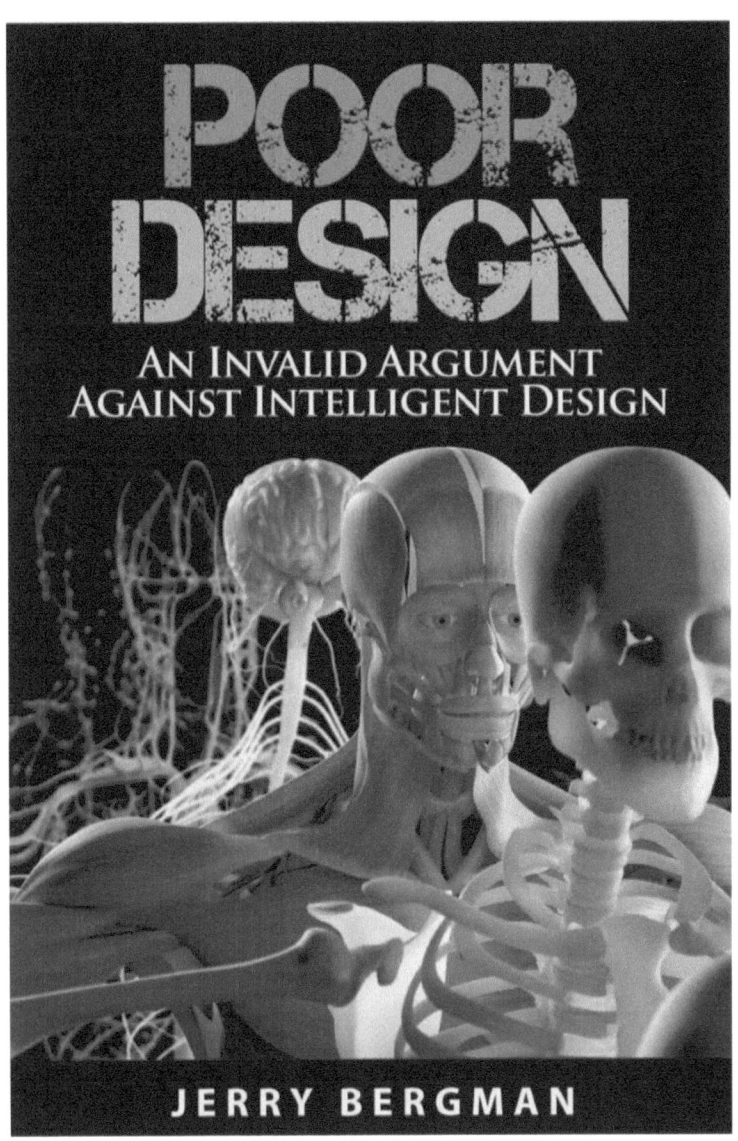

Claims of Poor Design Refute in detail by the author. The claim that the human body contains many examples of poor design is one of the strongest proofs of evolution. Author's book.

15

The Wielenberg Argument Against Darwinism from Good Design

ONE VERY POPULAR ARGUMENT for Darwinism, the argument from poor design, appears in the opening chapter of *The Problem of Pain*. Lewis countered with the following query: If the existing universe is so horribly designed

> how on earth did human beings ever come to attribute it to the activity of a wise and good Creator? Men are fools, perhaps; but hardly so foolish as that ... The spectacle of the universe as revealed by experience can never have been the ground of religion: it must always have been something in spite of which religion, acquired from a different source, was held.[1]

Professor Erik Wielenberg even claimed, "Lewis realizes that the design argument could never lead to a *good* Higher Power; in fact, insofar as it tells us anything about the moral attributes of the Higher

---

1. Lewis, 2001a, pp. 3-4.

Power at all, it points *away* from a good Power."[2] One example Hooper gives to support Wielenberg's interpretation is contained in a letter that Lewis wrote in 1946:

> The early loss of my mother, great unhappiness at school, and the shadow of the last war and presently the experience of it, had given me a very pessimistic view of existence. My atheism was based on it: and it still seems to me that *far* the strongest card in our enemies' [the atheists] hand is the actual course of the world ... I still think the argument from design the weakest possible ground for Theism, and what may be called the argument from un-design the strongest for Atheism.[3]

Beversluis adds that before Lewis became a Christian, the problem of pain "had been one of Lewis' chief objections to Christianity", and this fact may help to explain why he wrote the book *The Problem of Pain.*[4] He even opened the book, saying "not many years ago when I was an atheist...."[5] These concerns actually do not reflect a problem of the argument from poor physical design, but from the evil in the world (the loss of his mother, his unhappiness in school, the war and similar trials). Attempts to disprove the argument from physical design include the putative poor design of the human eye (its retina is backwards, and it has a blind spot) and the poor design in the recurrent lingual nerve, both claims have been refuted.[6] This is *not* the concern of Lewis here.

As is obvious in his discussion, Lewis' evidence *for* the design argument was primarily the moral argument. Specifically, Lewis believed

---

2. Quoted in Lewis, 2008, p. 118.
3. Lewis, 2008, p. 118.
4. Beversluis, John. 1985. "Beyond the Double Bolted Door." *Christian History* 4(3):29, July.
5. Hooper, Walter. 1996, p. 23.
6. Bergman, Jerry and Calkins, Joseph M.D. 2009. "Why the Inverted Human Retina is a Superior Design?" *CRSQ* 45(3):213-224, Winter.

that the best evidence of the existence of God is "moral phenomena" everywhere that proves "the existence of a Higher Power that created the universe. ...The Higher Power issues instructions and wants us to engage in morally right conduct. ...Then there is a good mind like the Higher Power that created the universe."[7]

Wielenberg, from his intensive study of Lewis, concluded that Lewis believed every person possesses a supernatural capacity, namely

> reason, which enables us to have genuine knowledge. Human reason must have some source, and since it has already been established that human reason cannot have been produced by nature, it must have a supernatural source. The supernatural source turns out to be God.[8]
>
> Human minds, then, are not the only supernatural entities that exist. They do not come from nowhere. Each has come into Nature from Supernatural: each has its tap-root in an eternal, self-existent, rational Being, whom we call God ... [H]uman thought is ... God-kindles.[9]

Thus, God designed humans to be moral, and Lewis' argument for God to support this view from reason is as follows:

1. If Naturalism is true, then knowledge exists only if natural selection could produce a capacity for knowledge starting with creatures with no such capacity.
2. But natural selection could not produce a capacity for knowledge starting with creatures with no such capacity.
3. So: If Naturalism is true, then knowledge does not exist (from 1 and 2).
4. But knowledge does exist.

---

7. Wielenberg, 2008, p. 63.
8. Wielenberg, 2008, p. 100.
9. Lewis, 1996c, Quoted from Erik J. Wielenberg, 2008, p. 99.

5. So: Naturalism is false (from 3 and 4).
6. If knowledge exists and Naturalism is false, then there is a supernatural, eternal, self-existent, rational Being that is the ultimate source of all knowledge.
7. Therefore, there is a supernatural, eternal, self-existent, rational Being that is the ultimate source of all knowledge (from 4, 5, and 6).[10]

In short, Lewis rejected evolution because he concluded "that it is 'not conceivable' that evolutionary processes could produce creatures capable of knowledge from creatures incapable of knowledge."[11] Wielenberg adds that one interpretation of this is "Lewis is claiming that he (and perhaps the reader as well—perhaps *everyone*) cannot conceive of any way in which evolutionary processes could produce beings capable of innate morality and knowledge. Let us say that when something is inconceivable in this sense, it is *weakly* inconceivable."[12]

Another interpretation that Wielenberg discusses is Lewis' claim "that he can see .... it is impossible for evolutionary processes to yield beings capable of knowledge" because "when something is inconceivable in this sense, it is *strongly* inconceivable."[13] An example is the concept of a round square

> is inconceivable, not merely in that I cannot conceive of a process that would produce such a shape (although that is true), but also in that I can see in a rather direct way that no such shape could exist ... When Lewis claims that the production of beings capable of knowledge by way of evolutionary processes is inconceivable, does he mean to say that it is weakly inconceivable or strongly inconceivable? ...It seems, therefore, that Lewis

---

10. Wielenberg, 2008, p. 101.
11. Wielenberg, 2008, p. 101.
12. Wielenberg, 2008, pp. 101-102.
13. Wielenberg, 2008, pp. 101-102.

must be claiming that he can see that the production of beings capable of knowledge by evolutionary processes is impossible. What support does Lewis offer for such a claim?[14]

Wielenberg concludes that the major problem Lewis had with the belief

that knowledge could arise via evolutionary processes is that he thinks it is impossible that intentional mental states could be created by evolutionary processes. Nature alone cannot produce intentionality; for this, you need something outside of nature, something that Lewis calls "reason." Without supernatural reason, there would be no thinkers capable of thinking *about* the natural universe.[15]

To document this, he quotes Lewis who notes that acts

of reasoning are not interlocked with the total interlocking system of Nature as all its other items are interlocked with one another. They are connected with it in a different way; as the understanding of a machine is certainly connected with the machine but not in the way the parts of the machine are connected with each other. The knowledge of a thing is not one of the thing's parts. In this sense something beyond Nature operates whenever we reason.[16]

Lewis here presents a complicated and challenging argument. The weakness of the argument includes the loophole that even though "the fact that evolutionary forces couldn't accomplish the task *in that particular way*, it does not follow that they couldn't accomplish it *at all.*"[17] Nonetheless, despite

---

14. Wielenberg, 2008, pp. 101-102.
15. Wielenberg, 2008, pp. 103-104.
16. Wielenberg, 2008, pp. 103-104.
17. Wielenberg, 2008, p. 104.

this weakness, the argument highlights a real puzzle for naturalism, and drawing attention to this puzzle is among Lewis' most important contributions to contemporary philosophy. By Lewis' own account, doubt about the compatibility of naturalism and knowledge was one of the main intellectual components of his abandonment of naturalism and eventual conversion to Christianity. Lewis credits his friend Owen Barfield withdrawing his attention to the difficulty.[18]

This line of reasoning has impressed many scholars, including the Notre Dame philosopher Alvin Plantinga who "proposed a much-discussed argument that owes much to Lewis' argument in *Miracles*."[19] On the other hand, in *Mere Christianity*, Lewis reasons that if we were forced

> to base our knowledge about God's nature exclusively on what we know of the observable physical universe, "we should have to conclude that the He was a great artist (for the universe is a very beautiful place), but also that He is quite merciless and no friend to man (for the universe is a very dangerous and terrifying place.")[20]

The latter is actually not an argument from classical design, but rather, as noted above, from the evil existing in the world, an undebatable fact. Conversely, most of the evil is caused by our fellow humans, not nature. The universe may be dangerous if we want to land on a star or another planet, but the only realistic danger we face are earthquakes, floods, hurricanes, and tornadoes, most of which can be dealt with by proper building construction and location concerns; for example, not building in a floodplain.

The agnostic Bertrand Russell himself "never wavered in his endorsement of the thesis that arguments from design cannot by them-

---

18. Wielenberg, 2008, p. 104.
19. Wielenberg, 2008, p. 105.
20. Wielenberg, 2008, pp. 181-182.

selves establish the existence of the traditional God of Christianity."[21] Rather, like both

> Hume and Lewis, Russell saw evil in the universe as one of the major stumbling blocks for such arguments. Many have pointed to Darwin's theory of evolution as putting a dagger through the heart of the argument from design, and in some places Russell endorses this view.[22]

C. S. Lewis, though, had much to say on this controversy. As noted in Chapter 10, evolution is based firmly on naturalism, a worldview that Lewis strongly and resoundingly rejected, writing that if "naturalism is true every finite thing or event must be (in principle) explicable in terms of Total Systems," admitting that "many things will only be explained when the sciences have made further progress." But he was not optimistic that this would ever occur.[23]

Conversely, Lewis had a problem with certain claims by scientists. For example, he found it "almost impossible to believe that the scientists really believe what they seem to be saying."[24] Evolution would be one idea included in the statement that "scientists really believe what they seem to be saying." I can hear Lewis asking, "do scientists really believe that all life resulted from time, chance, mutations and the outworking of natural law?"

Lewis' view of Genesis was explained in his book *Miracles*, writing "one seldom meets people who have grasped the existence of a supernatural God and yet deny that he is the Creator. All the evidence we have points in that direction, and difficulties spring up on every side if we try to believe otherwise." Lewis then writes that no

---

21. Wielenberg, 2008, p. 183.
22. Wielenberg, 2008, p. 183.
23. Lewis, 1996c, p. 17.
24. Lewis, 1996c, p. 20.

theory which I have yet come across is a radical improvement on the words of Genesis, that 'In the beginning God made Heaven and Earth.' I say radical improvement, because the story in Genesis—as St. Jerome said long ago—is told in the manner 'of a popular poet,' or as we should say, in the form of a folk tale. But if you compare it with the creation legends of other peoples—with all these delightful absurdities in which giants to be cut up and floods to be dried up are made to exist *before* creation—the depth and originality of this Hebrew tale will soon be apparent. The idea of *creation* in the rigorous sense of the word is there fully grasped.[25]

Lewis had a lot more to say about Genesis, but first we will look at Lewis' views on theistic evolution.

---

25. Lewis, 1996c, pp. 50-51. Emphasis in original.

Henri Bergson popularized one form of the Theistic Evolution form of Darwinism. His view was rejected in detail by Lewis.
From book cover.

16
Lewis Rejects the Theistic Evolutionism of Henri Bergson

As LEWIS CONTINUED his research, he increasingly found theistic evolution problematic. An example is Henri Bergson's (1859-1914) 'Creative Evolution' theory, a form of theistic evolution, specifically evolution directed by God.[1] Nonetheless, Bergson penned in detail one of the most effective set of critiques of Darwinian evolution ever. Most of his concerns are still very relevant today. Reading his writings no doubt greatly impressed Lewis, and thus he acknowledged that the Bergsonian critique of orthodox Darwinism was very impressive.[2]

Lewis first read Bergson's 1907 book titled *Creative Evolution* translated into English in 1911, "as a 19-year-old soldier during World War I while recovering from shrapnel wounds." Lewis agreed with Bergson's conclusion that the dynamic phenomena of life "cannot be contained

---

1. Bergman, Jerry. 2007. "Creative Evolution: An Anti-Darwin Theory Won a Nobel" *Impact* **409**:1-4, July.
2. Lewis, C. S. 1976. *The Weight of Glory*. New York: HarperCollins, p. 136.

within the mechanistic assumptions of nineteenth-century science," but disagreed with his attempt to weld theism with Darwinism.[3]

After reading Bergson, Lewis, although he agreed with Bergson's incisive criticisms of Darwinism, felt obligated to respond to Bergson's arguments that he disagreed with.[4] One example was Bergson's conclusion that "Life, from its beginnings, is the continuation of a single and identical drive (elan) which has divided itself among the divergent lines of evolution."

In order to respond to Bergson, Lewis added a section to his book *Mere Christianity* that was originally based on a radio manuscript he wrote. In *Mere Christianity*, Lewis mentioned only the materialist and religious view, so he felt it necessary to add a discussion of theistic evolution. He called the Life-Force philosophy of Bergson's *Creative Evolution* the "In-between view," or "Emergent Evolution" (which was often termed Darwinism in this book) because he concluded it was in-between atheism and theism. Lewis wrote in *Mere Christianity* that those who hold Bergson's view actually teach

> that the small variations by which life on this planet "evolved" from the lowest forms to Man were not due to chance but to the "striving" or "purposiveness" of a Life-Force. When people say this we must ask them whether by Life-Force they mean something with a mind .... If they do, then "a mind bringing life into existence and leading it to perfection" is really a God, and their view is thus identical with the Religious.[5]

He adds if they do not accept this Life-Force view of Bergson

> then what is the sense in saying that something without a mind "strives"

---

3. Schwartz, Sanford. 2009. *C. S. Lewis on the Final Frontier: Science and the Supernatural in the Space Trilogy.* New York: Oxford University Press, p. 89.

4. West, 2012, p. 125.

5. Lewis, 1980, p. 335.

or has "purposes"? This seems to me fatal to their view. One reason why many people find Creative Evolution so attractive is that it gives one much of the emotional comfort of believing in God and none of the less pleasant consequences.[6]

Of course, the same is true of other forms of theistic evolution. Bergson was "an unsparing critic of the creative power of Darwinian natural selection,"[7] firmly believing that natural selection could not explain the existence of the natural world.   Although Lewis strongly objected to Bergson's theistic evolution idea, he learned a great deal about the lethal problems of Darwinism from reading Bergson and the works of others. These ideas were expressed in his own writings. In fact, one of Lewis' most heavily annotated books was Bergson's nearly 400-page tome, *Creative Evolution,* that critiqued the creative ability of Darwinian natural selection theory.

Lewis also wrote many notes in Bergson's book, and underlined an even larger number of passages, later stating that the book's "critique of orthodox Darwinism" was very effective and that it "is not easy to answer."[8] Lewis concluded that the "Darwinian idea of adaptation by automatic elimination of the non-adapted is a simple and clear idea" because it attributes the cause to the environment outside of the organism, namely by natural selection, which controls evolution. The problem Lewis saw with this idea is natural selection is a negative, not a positive, influence. It subtracts, not adds, which is what evolution requires. Lewis admitted that he (Lewis) had a

> "great difficulty in accounting for the progressive and, so to say, rectilinear development of complex apparatus" like the vertebrate eye. Bergson stressed that Darwinism's reliance on accidental variation as the raw

---

6. Lewis, 1980, p. 335.

7. West, 2012, p. 125.

8. Lewis, 1996f, p. 89.

material for evolution made the development of highly coordinated and complex features found in biology nothing short of incredible. This was the case regardless of whether the accidental variations were slight or large.[9]

While Darwinians postulated that the mutations which occur in life are the source of variation that natural selection selects from, Bergson also noted that the Darwinians argued that these must have had very minor effects so as not to hinder the survival of the organism:

> For a difference which arises accidentally at one point of the visual apparatus, if it be very slight, will not hinder the functioning of the organ; and hence this first accidental variation can, in a sense, wait for complementary variations to accumulate and raise vision to a higher degree of perfection.[10]

Bergson observed that the many problems with this incorrect assumption of Darwinism, including the fact that

> the insensible variation does not hinder the functioning of the eye, neither does it help it, so long as the variations that are complementary do not occur. How, in that case, can the variation be retained by natural selection? Unwittingly one will reason as if the slight variation were a toothing stone set up by the organism and reserved for a later construction.[11]

Bergson noted that mutations are obviously of little comfort for Darwinism "which emphasizes that natural selection acts mechanically and without foresight."[12] Darwinists dealt with this problem by claiming "evolution relied on large accidental variations that provided

---

9. West, 2012, p. 125.
10. West, 2012, p. 126.
11. Bergson, Henri. 1944. *Creative Evolution*. New York: Random House, p. 72.
12. West, 2012, p. 125.

evolutionary leaps."¹³ This assumption creates "another problem, no less formidable" for Darwinism because, as Bergson asked, echoing the concept of irreducible complexity,

> how do all the parts of the visual apparatus, suddenly changed, remain so well coordinated that the eye continues to exercise its function? For the change of one part alone will make vision impossible, unless this change is absolutely infinitesimal. The parts must all change at once, each consulting the others.[14]

Bergson added that

> supposing chance to have granted this favor once, can we admit that it repeats the self-same favor in the course of the history of a species, so as to give rise, every time, all at once, to new complications marvelously regulated with reference to each other, and so related to former complications as to go further on in the same direction?[15]

In short, Bergson recognized that the "sheer improbability of the Darwinian explanation increases exponentially once one realizes how frequently the same complex biological features are supposed to have arisen independently in different evolutionary lineages."[16] Lewis concluded that the "subtlest foe of all—the Darwinians" was part of some plot, which Bergson exposed.[17]

In Bergson's words, "What likelihood is there that, by two entirely different series of accidents being added together, two entirely different evolutions will arrive at similar results?" We see this claim in the theory

---

13. West, 2012, p. 125.
14. Bergson, 1944, pp. 72-73.
15. Bergson, 1944, pp. 72-73.
16. West, 2012, p. 126.
17. C.S. Lewis Letter #1415 to Dr. Warfield M. Firor a professor of surgery at Johns Hopkins University December 20, 1951

evolutionists call *convergent evolution*. The idea of evolution, Bergson concluded, was incredible because an "accidental variation, however minute," requires many

> small physical and chemical causes. An accumulation of accidental variations, such as would be necessary to produce a complex structure, requires therefore the concurrence of an almost infinite number of infinitesimal causes. Why should these causes, entirely accidental, recur the same, and in the same order, at different points of space and time?[18]

Bergson's answer to his own question was that no informed, intelligent person could accept this theory of evolution "and the Darwinian himself will probably merely maintain that identical effects may arise from different causes, that more than one road leads to the same spot", adding we shall not

> be fooled by a metaphor. The place reached does not give the form of the road that leads there; while an organic structure is just the accumulation of those small differences which evolution has had to go through in order to achieve it.[19]

Consequently, the "struggle for life and natural selection can be of no use to us in solving this part of the problem, for we are not concerned here with what has perished, ... only with what has survived." From the "extensive annotations Lewis made in his personal copy of *L'Evolution Cré*atice, it is clear that he understood and greatly appreciated Bergson's critique of" Darwinism.[20] West writes that, furthermore

> Lewis aptly summarized the Darwinian mechanism of adaptation according to Bergson as the "[e]limination of the unfit" and noted that it "plainly cannot account for complicated similarities on divergent lines

---

18. Bergson, 1944, p. 74.
19. Bergson, 1944, p. 74.
20. West, 2012, p. 127.

of evolution." Lewis also noted Bergson's view that "pure Darwinism has to lean on a marvelous series of accidents" and how Darwinists try to "escape" this truth "by a bad metaphor." Lewis paid particular attention to Bergson's critique of Darwinian accounts of eye evolution in mollusks and vertebrates, concluding that "[n]atural selection... fails to explain these eyes."[21]

The acceptance of Bergson's very effective concise critique of Darwinism is indicated by his being awarded the most prestigious honor in academia, the 1927 Nobel Prize, for "achievements that have conferred the greatest benefit to humankind." The prize added that it was "in recognition of his rich and vitalizing ideas and the brilliant skill with which they have been presented."[22] The French government also honored Bergson as late as 1959 on a postage stamp. One of the most biting comments made by Lewis about Bergson was that his "Creative Evolution is 'the religion of the Twentieth Century'"[23] because, although Bergson recognized evolution as untenable, it fell far short of Biblical creation.

---

21. West, 2012, p. 127.

22. https://www.nobelprize.org/prizes/literature/1927/bergson/facts/.

23. Lewis, 1967a p. 300.

C. S. Lewis strongly opposed Catholic Pierre Teilhard de Chardin's theistic evolution.

17

Lewis Opposes the Theistic Evolutionism of Pierre Teilhard de Chardin

LEWIS HAD LITTLE RESPECT for attempts to combine evolution and Christianity such as that attempted by Jesuit priest Pierre Teilhard de Chardin. Many years after he wrote the letter to his father dated August 14, 1925, explaining his conclusion that Charles Darwin's theory and Herbert Spencer's Social Darwinism theory were both built "on a foundation of sand," on March 5, 1960, he wrote that Jesuit priest Pierre Teilhard's mystical theistic evolution theory "which is being praised to the skies... is evolution run mad."[1]

Lewis also opined that the "Jesuits were quite right in forbidding him [Teilhard] to publish any more books" on his brand of theistic

---

1. Lewis, 2007, p. 1137; Ferngren, Gary B. and Ronald L. Numbers. 1996. "C. S. Lewis on Creation and Evolution: The Acworth Letters, 1944-1960." *The American Scientific Affiliation* **48**:28-33, March, p. 30.

evolution, adding that this "prohibition probably explains the success of de Chardin among scientists."² Teilhard (1881-1955) was a famous paleontologist as well as a Jesuit priest that the Church believed went far beyond Catholic orthodoxy in his philosophical writings and their publication was condemned.

Lewis even wrote to his friend, Father Frederick Joseph Adelmann, on September 21, 1960, opining that the Catholic Church was "right... to shut up de Chardin", adding that he [Lewis] was concerned about the "enormous boosts now being given to all that Bergsonian—Shavion—pantheistic—bioidolatrous waffling."³ Bioidolatrous referred to evolution worship.

Father Adelmann earned his Ph.D. from St. Louis University and chaired the philosophy department at the Catholic Boston College. Although Lewis attacked Teilhard's mystical theistic evolutionism as recounted in Teilhard's book *The Phenomenon of Man* written in 1938 and published only in 1955⁴ and not theistic evolution *per se*, his criticism nonetheless applies to other attempts to support theistic evolution.

Lewis regarded Teilhard's theistic evolution as "a unique theory of theistic evolution," a "spiritualized version of evolutionary development," where Christ is the initiator, the energizer, and the final end of the cosmic evolutionary process.⁵ Nonetheless, it was a theistic evolution theory that was actually not much different from those taught by

---

2. Lewis, 2007, p. 1137.
3. Lewis, 2007, p. 1186.
4. Lewis, 1970, p. 523.
5. Lewis, Gordon R. and Bruce A. Demarest. 2010. *Integrative Theology*. Grand Rapids, MI: Zondervan, p. 204; Lewis, C. S. 1996. *The Collected Works of C. S. Lewis: The Pilgrim's Regress, Christian Reflections, God in the Dock*. New York: Inspirational Press.

many modern theistic evolutionists. Teilhard de Chardin taught what was actually a vitalism idea that he called the "Omega Point", which was a *maximum* level of complexity and consciousness towards which, he believed, the universe was evolving. The mechanism behind this progression to the Omega point was never explained.

One example that illustrates Lewis' attitude toward both Darwinism and Teilhard's theistic evolution was provided by Professor Alister Fowler. Fowler wrote to Lewis in 1961 about the interesting ideas in Teilhard de Chardin." Lewis replied "accusing me, at least half seriously, of "biolatry": "You talk of Evolution as if it were a substance (like individual organisms) and even a rational substance or person. I had thought it was an abstract noun."[6]

Lewis was concerned about what he called the "Idol of Evolution" in de Chardin's teaching because "his organizing key was not merely biological evolution, but an expanded, cosmic evolution of the universe."[7] Teilhard de Chardin taught that evolutionary progress requires humanity's cooperation, implying a eugenics involvement in his "spiritualized version of evolutionary development."[8]

One of Lewis' other major reasons for opposing Pierre Teilhard de Chardin's theistic evolution was because it was largely only "a restatement of Bergson"[9] who, as documented in Chapter 16, Lewis also opposed. Lewis explains that among the reasons Bergson's theistic evolution theory was very attractive to many persons was because it had none of the negative consequences of Christian theism. As Lewis wrote, during good times when "you do not want to believe that the whole universe is a mere mechanical dance of atoms, it is nice to be

---

6. Fowler, 2003, p. 71. Letter to Fowler May 4, 1961.
7. Lewis and Demarest, 2010, p. 204.
8. Lewis and Demarest, 2010, p. 204.
9. Lewis, 2007, p. 1257.

able to think of this great mysterious Force rolling on through the centuries and carrying you on its crest."[10] Conversely, if

> you want to do something rather shabby, the Life-Force, being only a blind force, with no morals and no mind, will never interfere with you like that troublesome God we learned about when we were children. The Life-Force is a sort of tame God. You can switch it on when you want, but it will not bother you. All the thrills of religion and none of the cost. Is the Life-Force the greatest achievement of wishful thinking the world has yet seen?[11]

Lewis reveals his opposition to theistic evolution with his statement that it "is not Christianity which need fear the giant universe," rather, it is those belief systems that attempt to explain the "whole meaning of existence" by

> biological or social evolution on our own planet. It is the creative evolutionist, the Bergsonian or Shavian ... who should tremble when he looks up at the night sky. For he really is committed to a sinking ship ... attempting to ignore the discovered nature of things, as though by concentrating on the possibly upward trend in a single planet he could make himself forget the inevitable downward trend in the universe as a whole, the trend to low temperatures and irrevocable disorganization. For *entropy is the real cosmic wave, and evolution only a momentary tellurian ripple within it.*[12]

Lewis correctly noted the influence of Bergson on Teilhard. Teilhard studied theology at Hastings in Sussex, from 1908 to 1912. It was there that he synthesized his scientific, philosophical and theological knowledge in harmony with evolution. It was at this time that Teilhard read *L'Évolution Créatrice* (The Creative Evolution) by a fellow

---

10. Lewis, 1980, pp. 34-35.
11. Lewis, 1980, pp. 34-35.
12. Lewis, 1970, p. 44. Emphasis added.

Frenchman, Henri Bergson. Lewis wrote that when he read Bergson's book "the only effect that brilliant book had on me was to provide fuel at just the right moment... for a fire that was already consuming my heart and mind."[13] This fire was to aggressively oppose Darwinism.

Teilhard added the importance of "that magic word 'evolution' which haunted my thoughts like a [musical] tune: which was to me like an unsatisfied hunger, like a promise held out to me, like a summons to be answered."[14] In short, intoxicated by evolution, his theory was Darwinism covered with a thin layer of theism. Bergson's ideas helped Teilhard to unify his beliefs on matter, life, and energy into what he concluded was a coherent worldview.

To blend the two, Teilhard had to distort both evolutionism and creationism.[15] Thus, he created a new theistic theory of evolution which, according to devastating reviews by leading mainline scientists of his masterpiece *The Phenomenon of Man*, failed miserably in the attempt. Critics believe he distorted both science and religion. Nobel Laureate Peter Medawar wrote that *The Phenomenon of Man* "was nothing but non-sense."[16] In the end, Lewis concluded that "worship and service of evolution, rather than the transcendent Lord of all, is idolatrous."[17]

Teilhard's ideas still resonate with some today, especially New-Age adherents. Nevertheless, their acceptance has greatly diminished in the last decade or so, as evidenced by the demise of the magazine *The Teil-*

---

13. Teilhard de Chardin, Pierre. 1979. *The Heart of Matter*. New York: Harcourt Brace Jovanovich, p. 25. (Translated from the French by René Hague.)
14. Teilhard de Chardin, 1979, p. 25.
15. Delfgaauw, Bermard. 1969. *Evolution: The Theory of Teilhard de Chardin*. New York: Harper & Row.
16. Schumacher, Leo S. 1968. *The Truth About Teilhard*. New York: Twin Circle Publishing Company, p. 8.
17. Lewis and Demarest, 2010, p. 204.

hard Review, published three times a year from 1966 to 1981 by the *Teilhard Center for the Future of Man* in London. Other Teilhard organizations included *The American Teilhard Association*, and additional allied groups in Germany, France, Italy and even Japan.[18]

Lewis opposed Teilhard because of his heretical, anti-Christian ideas, such as the Omega point. However, from my review of the articles in *The Teilhard Review*, which I subscribed to since the first issue in 1966, the publication included many articles supporting Intelligent Design and opposing Darwin, but not evolution, such as "Darwin & the Death of Natural Theology" by Zoology Professor Mathison at North East London Polytechnic. This article discusses favorably the importance of William Paley, writing that God is not only the Divine Designer but a feudal Prince and Intercessor."[19]

Professor Mathison concluded that "The theory of evolution ... displaced the concept of God as a mechanic and replaced it by a vacuum or a contradiction."[20] This new paradigm Teilhard attempted requires filling in this vacuum by merging natural theology with an evolutionary worldview. As a result, the "conception of evolution permeates Teilhard's writings, whether on philosophy, metaphysics, theology or biology."[21] No doubt for this reason, Lewis took umbrage at Teilhard's philosophy. This is also why the "new consciousness movements frequently quote Teilhard de Chardin's spiritualized version of

---

18. The American Teilhard Association. http://teilharddechardin.org/index.php/forerunners-in-europe; https://www.teilhard.org.uk/resources/allied-organisations/.

19. Mathison, Jane. 1979. "Darwin & the Death of Natural Theology." *The Teilhard Review* **14**(1):23-35.

20. Mathison, 1979, p. 28.

21. Mathison, 1979, p. 30.

evolutionary development."[22]

After repeating the Stanley Miller abiogenesis view of the origins of life, Professor George Bishop concluded that the "simplest and most scientific explanation is to admit life must have been created by ONE able to produce life from inert matter… To bring us into existence, it has from the beginning juggled miraculously with too many improbabilities… Teilhard conceived of evolution as created and guided by God."[23] Thus, evolution attempts to tell the entire story of the origins of the Earth, life, and the universe, but evolution has never been able to fill in the enormous gaps, even by speculative just-so stories. The gaps include between life and non-life, between asexual life and sexual life, between prokaryotes and eukaryotes, and between invertebrates and vertebrates. Again, in Teilhard's worldview, evolution was at the core of all creation. No wonder Lewis was so strongly opposed to Teilhard.

Lewis regarded the question of the origins of both life and the universe as critically important. As evidence of this, he observed that "since men were able to think, they have been wondering" how our universe and life began.[24] He then explained that two main views have existed about origins, first

> there is what is called the materialist view. People who take that view think that matter and space just happen to exist, and always have existed, nobody knows why; and that the matter, behaving in certain fixed ways, has just happened, by a sort of fluke, to produce creatures like ourselves who are able to think.[25]

Lewis was obviously unimpressed with this view, called naturalism

---

22. Lewis and Demarest, 2010, p. 204.
23. Bishop, George. 1980. Evolution: "Blind Chance or God?" *The Teilhard Review* **15**(2):17-23, p. 23.
24. Lewis, 1980, p. 31.
25. Lewis, 1980, p. 31.

today, explaining, in harmony with the Intelligent Design view, that the probability is less than one in a thousand that something could have collided with the Sun, causing it to

> produce the planets; and by another thousandth chance the chemicals necessary for life, and the right temperature, occurred on one of these planets, and so some of the matter on this earth came alive; and then, by a very long series of chances, the living creatures developed into things like us.[26]

Lewis explained that the Intelligent Design worldview is far more reasonable because "what is behind the universe is more like a mind than it is like anything else we know ... it is conscious, and has purposes, and prefers one thing to another ... it made the universe ... to produce creatures like itself ... to the extent of having minds."[27]

Lewis then attacked the view that creationism was superseded by the modern, "scientific" view, namely evolution. He did not agree with the scientists who concluded that evolution replaced the creation view that has been held since Old Testament times. One reason Lewis gave for why the creation view has been, and still is, accepted by many scientists is the fact that "Wherever there have been thinking men both views [creation and evolution] turn up."[28]

Lastly, Lewis stressed that science all too often has become a religion, and one cannot determine which view, creation or evolution, is the correct view of origins by science alone because science proper is the process of gaining knowledge

> by experiments. It watches how things behave. Every scientific statement in the long run, however complicated it looks, really means something

---

26. Lewis, 1980, p. 32.

27. Lewis, 1980, p. 32.

28. Lewis, 1980, p. 32.

like ... "I put some of this stuff in a pot and heated it to such-and-such a temperature and it did so-and-so."[29]

Lewis then stressed that he is not degrading science in general, but rather is only saying what the job of science is, adding that "the more scientific a man is, the more [I believe] he would agree with me that this is the job of science—and a very useful and necessary job it is ... But why anything comes to be ... and whether there is anything behind the things science observes ... is not a scientific question" but a religious or, at the least, a worldview question.[30]

Lewis concluded that the origins-of-life problem is not a question that the scientific method can answer because, if there is "Something Behind" the creation, "then either it will have to remain altogether unknown to men or else make itself known in some different way ... And real scientists do not usually make [claims about theology].... It is usually the journalists and popular novelists who have picked up a few odds and ends of half-baked science from textbooks who [do]."[31]

In other words, Lewis did not have a problem with the results of empirical research supporting microevolution, the process that creationists refer to as variation within the Genesis kinds. Modern examples of this work by scientists includes research on bacterial resistance, the peppered moth, small variations in the size of Darwin's finch beaks and the like. In his later writing, Lewis detailed in more depth his concern about both macroevolution and naturalism.

Lewis repeatedly stressed that science has its place, but it also has clear limits, writing: "Let scientists tell us about sciences. But government involves questions about the good for man, and justice, and what things are worth having at what price; and on these [things] scientific

---

29. Lewis, 1980, p. 32.
30. Lewis, 1980, p. 32.
31. Lewis, 1980, p. 32.

training gives a man's opinion no added value."³² Canadian journalist Denyse O'Leary concluded from her study of Lewis that he was

> careful to distinguish between evolution as a theory in biology and Evolution as an idea that came to dominate the politics and religion of his time. He noted that decades before Darwin's *On the Origin of Species,* poets and musicians had started proclaiming that humanity was inevitably evolving, onward and upward, to a glorious future. "I grew up believing in this Myth and I have felt—I still feel—its almost perfect grandeur."³³

She added that Lewis believed a critical difference existed

> between the myth of Evolution preached by poets and the theory in biology. Lewis made quite clear to his readers that the biological theory … does not argue that there will be inevitable, continuous *improvements* over time. It only explains continuous *change* over time. The change can be degeneration or decline…as applied to human society, the myth of evolution morphed from a *theory* about changes to a supposed *fact* about improvements. It became a powerful weapon for good or ill in the hands of social reformers and politicians.³⁴

O'Leary here refers to the problem that entropy, the tendency for deterioration and disorder to occur in the natural world, presents for evolution.³⁵ In his "The Funeral of a Great Myth" essay, Lewis wrote that, for persons who were "brought up on the Myth [of evolution], nothing seems more normal, more natural or more plausible than that chaos should turn into order, death into life, ignorance into knowl-

---

32. Lewis, 1970, p. 315.
33. O'Leary, Denyse. 2004. *By Design or by Chance?* Ontario, Canada: Castle Quay Books, p. 63.
34. O'Leary, 2004, pp. 63-64.
35. See Lewis, 1970, p. 44.

edge."[36]

Of course, Lewis recognized that chaos will not turn into order without an orderer, and death (meaning nonliving material such as minerals) will not turn into life without a life-giver, and that is clearly his point! O'Leary adds that Lewis believed "continuous improvement without human effort seemed so plausible" to Darwinists and

> that contrary examples in nature were simply ignored. For example, popular presentations of evolution typically feature the evolution of the modern horse from a bulgy little pawed creature. They do not highlight the evolution of an active, independent creature into a degenerate, disease-causing parasite, though that happens, too.[37]

Lewis also had some sharp comments to make about the realities of science. In his largely autobiographical *The Pilgrim's Regress*, one of the first books he published after his conversion to Christianity,, he debunked modern arrogant unbelief based on evolution and guesswork which masquerades as "science." Lewis also made it clear that this pseudoscience is what he had in mind when he referred to *emergent evolutionism* or developmentalism; what is today called "Darwinism" or "gradualism."

In *The Pilgrim's Regress*—which follows the progress of a fictional character through the philosophical landscape, including nihilism, before eventually arriving at traditional Christianity—Lewis wrote that the lead character in his book, as is true of so many persons today, had "a very crude notion of how science actually works." Lewis put the following words in the mouth of the foolish old "Mr. Enlightenment," namely, that a scientific hypothesis

---

36. C. S. Lewis quoted in O'Leary, 2004, p. 112.

37. O'Leary, 2004, p. 112.

establishes itself by a cumulative process: or, to use popular language, *if you make the same guess often enough it ceases to be a guess and becomes a Scientific Fact.*

After he had thought for a while, John said:

'I think I see. Most of the stories about the Landlord are probably untrue; therefore the rest are probably untrue.'

'Well, that is as near as a beginner can get to it.... But when you have had ... scientific training you will find that you can be quite certain about all sorts of things which now seem to you only probable.'[38]

How appropriate these sarcastic words are now, so many years later, about evolutionary dogma. In contrast to the overwhelming evidence to the contrary, many writers still claim that Lewis was a committed theistic evolutionist.

Burson and Walls reason that the problem of evil should not be a "problem for those who do not believe in God. If matter and energy are ultimately all that exists, and *human beings arrived on the scene by a blind evolution* [process], there seems to be little reason to complain about evil and suffering… If atoms and energy are ultimately reality, … They do not hear our protests. They do not hear and they do not care."[39] This observation was one reason Lewis had a major problem with evolution, as he repeatedly made clear in his writings.

Summary

History has not been very kind to the ideas and person of Teilhard. Controversy even exists on his attempts to prove human evolution. Harvard Professor Stephen Gould attempted to document Teilhard de

---

38. Lewis, C. S. 1992. *The Pilgrim's Regress: An Allegorical Apology for Christianity, Reason, and Romanticism*. Grand Rapids, MI: Eerdmans, p. 28. Emphasis added.
39. Burson and Walls, 1998, p. 200. Emphasis added.

Chardin's involvement in the most famous fraud in history, the Piltdown affair,[40] a charge that the American Teilhard Association has attempted to refute.[41] Teilhard was also involved in the Peking Man fossil mystery (*Sinanthropus pekinensis*) which mysteriously disappeared in 1941 in China. Although casts still exist, they do not substitute for the real bones.[42]

Peking Man is now considered a modern human, thus renamed *Homo erectus pekinensis*. Discovered in 1929–1937 during excavations at Zhoukoudian near Beijing (at that time named Peking, thus its name), most of the early studies of the fossils were conducted by Davidson Black until his death in 1934. Pierre Teilhard de Chardin took over (until Franz Weidenreich replaced him) and studied the fossils until he left China in 1941.[43]

---

40. Gould, Stephen Jay. 1980. *The Panda's Thumb*. Chapter 10: "Piltdown Revisited", pp. 108-124. New York: W.W. Norton & Company.

41. McCulloch, Winifred. 1996. *Teilhard de Chardin and the Piltdown Hoax*. Teilhard Studies Number 33. Lewisburg, PA: Bucknell University. http://teilharddechardin.org/old/studies/33-Teilhard_and_the_Piltdown_Hoax.pdf

42. Aczel, Amir. 2007. *The Jesuit and the Skull: Teilhard de Chardin, Evolution, and the Search for Peking Man*. New York: Riverhead Hardcover.

43. Janus, Christopher G. and William Brashler, 1975. *The Search for Peking Man*. New York: Macmillan.

Herman Muller who discover that x-rays cause mutations which causes genetic changes. For this discovery he received the Nobel Prize because mutations were believed to "speed up evolution." It is now known that the vast majority of mutations are harmful, and some are lethal.

18

Lewis Opposes the Foundation of Evolution

A MAJOR CONCERN Lewis had with evolution was the foundation of orthodox Darwinism – natural selection of mutations. This is the major mechanism that was theorized to drive evolution. Darwin observed that all life varies, even within a specific life-form. The types and variations of dogs, for example, include their color, size, ability to run and visual ability. Evolution theory requires an enormous amount of variety exists in the living world.

Of that variety, the theory postulates, some traits improve the chances of an animal's survival. Those animals that are better able to survive are more likely to live longer and, consequently, will have more offspring. As a result, they will gradually numerically out-reproduce those animals lacking that trait. This was Darwin's major contribution to evolution. Ironically, in the continuing absence of corroborating evidence, Darwin later questioned the ability of his natural selection mechanism to produce all of the millions of forms of life existing on Earth today.

In short, life-forms possessing variations that aid their survival are more apt to survive and thus reproduce. Life-forms with variations that put them at a survival disadvantage are less apt to survive, thus less likely to reproduce. The theory sounds logical and its logic is easy to follow. The problem is *the real world* is not nearly so simple. Often survival is more a matter of chance and luck. The sick and lame are more likely to be eaten or die prematurely, thus less likely to survive and reproduce. The result is natural selection often *reduces* weaker variants, thereby helping to stabilize populations.

This idea of natural selection was problematic for Lewis because he reasoned it violated the fact which the "incurably evolutionary or developmental character of modern thought is always urging us to forget [namely what] is vital and healthy does not necessarily survive. Higher organisms as apes are often killed by lower ones. Ants as well as men are subject to accident and violent death."[1]

Lewis later added his conclusion that "Darwinism gives no support to the belief that natural selection, working upon chance variations, has a general tendency to produce improvements."[2] He inferred that the claim that Darwinian natural selection was the cause of any kind of evolution was an illusion that results from observing only a few species that have been improved for human use by artificial breeding. Scientists, he noted, have unscientifically extended this limited observation to the whole of creation.[3] Lewis concluded that, contrary to Darwinism, "there is no general law of progress in biological history."[4]

Another reason Lewis opposed natural selection as the creator of

---

1. Lewis, C. S. 1954. *English Literature in the Sixteenth Century.* New York: Clarendon Press, p. 113.
2. Lewis, C. S. 1960. *The World's Last Night and Other Essays.* New York: Harcourt Brace Jovanovich, p. 103.
3. Lewis, 1960, p. 103.
4. Lewis, 1960, p. 103.

all life forms, a process termed by Herbert Spencer as "the Darwinian 'struggle for existence", was because it reduces man from a spiritual [being] to a purely biological entity [and] elevates the state of inexorable competition, conflict, and self-serving aggression from a tragic lapse of our ordained destiny into the primary principal of the natural order."[5] Focusing on human evolution, Lewis recognized that "natural selection could operate only by eliminating responses that were biologically hurtful and multiplying those which tended to [aid] survival. But it is not conceivable that any improvement of responses could ever turn them into acts of insight or even remotely tend to do so."[6]

This observation is true not only of human thoughts, but also other traits which will likely not result in species evolution. Writing that although natural selection may maximize one specific trait, even if our "psychological responses to our environment … could be indefinitely improved (from the biological point of view) without becoming anything more than responses,"[7] it was not readily apparent to Lewis how such envisioned *psychological* progress could translate into an organism's *physiological* survival benefit.

Anticipating how evolutionists might respond to his objections to natural selection, Lewis writes "perhaps we cannot exactly see—not yet—how natural selection would turn sub-rational [Lewis' term for non-rational] mental behavior into inferences that reach truth. But [evolutionists claim] we are certain that this in fact happened." Evolutionists "know" that evolution has happened only because we know that we are here, and the only explanation that they will accept is a naturalistic one, thus evolution.

This rationalization response is seen everywhere where evolution-

---

5. Lewis, 2001a, p. 28.
6. Lewis, 2001a, p. 28.
7. Lewis, 2001a, p. 29.

ists attempt to defend their worldview, such as the origin of sexual reproduction. They argue that, although we have no evidence that sex could evolve purely by the accumulations of mutations and natural selection, we know it did because we *know* evolution is true, therefore sexual reproduction must somehow have evolved from asexual reproduction.[8] We haven't yet figured out how it happened, and may never figure out how it happened, but we *know* it happened because the 'God explanation' is not allowed.

Lewis recognized that the description of the origin "of thought as an evolutionary phenomenon" was very problematic. He realized the difficulty evolutionists had in explaining the origin of humanity's higher level of thinking by mindless evolutionary mechanisms. Lewis reasoned that evolutionists make a "tacit exception in favor of [their view of] … thinking" that Lewis presents because evolution, i.e., Nature, is a process that is "quite powerless to produce rational thought; not that she never modifies our thinking, but that the moment she does so, it ceases (for that very reason) to be rational" because rational thinking requires conscious consideration of an issue based on one's past knowledge and values.[9]

---

8. Smith, F. LaGard. 2018. *Darwin's Secret Sex Problem: Exposing Evolution's Fatal Flaw—the Origin of Sex.* Bloomington, IN: WestBow Press.

9. Lewis, 2001a, pp. 36, 38.

Arthur C. Clark English science fiction and science writer, futurist, inventor, undersea explorer, and television series host.

19

Lewis Anticipated Arguments Used by Modern Evolutionists

LEWIS APPEARS TO HAVE anticipated some of the main arguments used by evolutionists today. One example is the natural selection of mutations, and he attempted to respond to them. One of the most popular arguments among Darwinists today is the Pale Blue Dot idea popularized by the late Carl Sagan. In short, Sagan argued that astronomical research has found

> that we live on an insignificant planet of a humdrum star lost between two spiral arms in the outside of a galaxy tucked away in some forgotten corner of a universe in which there are far more galaxies than people.[1]

Lewis writes the following challenge to this idea in one of his papers that was originally part of a lecture he presented in 1945. The lecture presented to a university audience raised the following claim: "Now that we know how huge the universe is and how insignificant the

---

1. Sagan, Carl. 1980. *Cosmos*. New York: Random House, p. 193.

Earth [is], it is ridiculous to believe that the universal God should be specially interested in our concerns." [2]

Lewis answered this claim by noting God may be very concerned with many parts of the universe, likely the entire universe, but our corner may be a special concern. Lewis correctly challenged the identification of size with importance, writing: "Is an elephant more important than a man, or a man's leg than his brain" based on size?[3] This issue was the focus of his dialogue with science writer Arthur C. Clarke.

Correspondence with Arthur Clarke

Lewis' correspondence was with leading British science-fiction and science writer, inventor, undersea explorer, and television series host Arthur C. Clarke (1917-2008). The dialogue helps us to understand the impressions his fellow science-fiction writers had of Lewis' views of Darwinism. Clarke also co-wrote the screenplay for the 1968 film *2001: A Space Odyssey*, a film claimed by some to be one of the most influential films of all time.

The film garnered a cult following and was the highest-grossing North American film of 1968. It was nominated for four Academy Awards and prominently featured evolution in a scene showing two tribes of pre-human apes fighting. In a eureka moment, one tribe discovered that they could effectively kill members of the other tribe by using bones of dead apes as weapons. After eliminating the other 'inferior' tribe, the film implies that the conquering, weapon-wielding tribe eventually evolved into modern humans.

Clarke "didn't believe in a personal god, but was open to suggestions about what caused the big bang." He was once a "physics student who wrote about evolution," and in his writing aggressively argued

---

2. Lewis, 1970.

3. Lewis, 1970.

for evolution and criticized religion.[4] As expected, traveling into outer space for "Clarke was potentially an evolutionary step forward. In one of Clarke's novels, some extraterrestrials come to Earth to somehow help us along in the evolutionary process. Conversely, Lewis was not only a skeptic of evolution, but was concerned about the potential harm we could do on another planet if we were able to travel there.[5]

In short, in contrast to Clarke, Lewis saw the human race, and scientists in particular, as terribly flawed, and even somewhat warlike. He also saw that the exploration of outer space might end up repeating our mistakes of the past, such as expanding imperialism into other areas of the universe. Lewis also wrote that the sole reason the Church exists is to draw men to Christ. And if the churches

> are not doing that, all the cathedrals, clergy, missions, sermons, even the Bible itself, are simply a waste of time. God became Man for no other purpose. It is even doubtful ... whether the whole universe was created for any other purpose. It says in the Bible that the whole universe was made for Christ ... We do not know what (if anything) lives in the parts of it that are millions of miles away from this Earth. Even on this Earth we do not know how it applies to things other than men... We have been shown the plan only in so far as it concerns ourselves.[6]

He even wondered if Nature, which he defined as actually only "space and time and matter," could ever produce a situation that was able to create every life-form in existence, not only today, but also in the past. For example, he asked, is there any

> other way of getting many eternal spirits except by first making many natural creatures, in a universe, and then spiritualizing them? ... The idea that the whole human race is, in a sense, one thing—one huge organism,

---

4. Miller, 2003, p. 7.
5. Miller, 2003, pp. 17-18.
6. Lewis, 1980, p. 171.

like a tree—must not be confused with the idea that individual differences do not matter or that real people ... are somehow less important than collective things like classes, races, and so forth. Indeed the two ideas are opposites.[7]

Lewis believed that God, through Christ, "created the whole universe, became not only a man but (before that) a baby, and before that a *fœtus* inside a Woman's body. If you want to get the hang of it, think how you would like to become a slug or a crab."[8]

Furthermore, he wrote that Christianity teaches that God not only created the world, but also that

space and time, heat and cold, and all the colors and tastes, and all the animals and vegetables, are things that God "made up out of His head" as a man makes up a story. But ... [the atheists] also think that a great many things have gone wrong with the world that God made and that God insists, and insists very loudly, on our putting them right again.[9]

This reasoning raises a major question, namely if

a good God made the world why has it gone wrong? And for many years I simply refused to listen to the Christian answers to this question, because I kept on feeling ... however clever your arguments are, isn't it much simpler and easier to say that the world was not made by any intelligent power? Aren't all ... arguments simply a complicated attempt to avoid the obvious?[10]

Of course, to Lewis the obvious answer is that a good God made the world good, and it was humans that corrupted it.

---

7. Lewis, 1980, p. 171.
8. Lewis, 1980, p. 156.
9. Lewis, 1980, p. 45.
10. Lewis, 1980, p. 45.

Clarke mentioned Lewis in an article titled 'Armchair Astronauts' in *Holiday Magazine* claiming, "Less sympathetic to our aims was Dr. C. S. Lewis, author of two of the very few works of space fiction that can be classed as literature – '*Out of the Silent Planet*' and '*Perelandra.*' Both of these fine books contained attacks on scientists in general, and astronauts in particular, which aroused my ire."[11] As already covered in detail, Lewis was not attacking science, or the scientific method, but rather scientists who went well beyond the science and used their authority to push their Darwinian worldview. Nowadays they use the educational system, the mainstream media, and the courts to ensure exclusive entrenchment and enforcement of their worldview, and *only* their worldview propaganda, to both students and in society.

---

11. Arthur C. Clarke & C. S. Lewis. http://oxfordinklings.blogspot.com/2007/06/arthur-c-clarke-cs-lewis.html. [Website sometimes claims that this blog page does not exist.]

Lewis objected to Albert Einstein's impersonal god. Lewis' picture of God was as a personal, Creator God. The Einstein picture was taken in 1933.

20

Lewis Teaches a Creation Worldview

LEWIS REPEATEDLY REJECTED Naturalism throughout his life after becoming a Christian. He not only criticized evolution, he also consistently taught the creation worldview in his various writings. This can be seen in his book *The Magician's Nephew*, the sixth of the *Chronicles of Narnia* novels. Narnia was a fantasy world created by C.S. Lewis as the primary location for his series of seven fantasy novels. The Narnia creation is described in ways that closely parallel the Biblical creation account.

Lewis also objected to Einstein's god because Lewis' picture of God was a personal, Creator God as taught by the Judeo-Christian tradition. He rejected the mechanism of emergent evolution, today often called atheistic evolution or macroevolution, because, like all materialistic systems, it

> breaks down at the problem of knowledge. If thought is the ... irrelevant product of cerebral motions, what reason have we to trust it? As for emergent evolution, if anyone insists on using the word *God* to mean 'whatever the universe happens to be going to do next', of course we

cannot prevent him. But nobody would in fact so use it unless he had a secret belief that what is coming next will be an improvement [which is emergent evolution].¹

Lewis added that this emergent evolution belief is not only unwarranted, but

> presents peculiar difficulties to an emergent evolutionist. If things can improve, this means that there must be some absolute standard of good above and outside the cosmic process to which that process can approximate. There is no sense in talking of 'becoming better' if better means simply 'what we are becoming'— it is like congratulating yourself on reaching your destination and defining destination as 'the place you have reached.' Mellontolatry, or the worship of the future, is a *fuddled* religion.²

Lewis has also thought about the implications of creationism, writing that

> God, besides being the Great Creator, is the Tragic Redeemer. Perhaps the Tragic Creator too. For I am not sure that the great canyon of anguish which lies across our lives is *solely* due to some pre-historic catastrophe. Something tragic may… be inherent in the very act of creation.³

Lewis also accepted the creation *ex nihilo* position, adding that when he speaks of God "uttering" or "inventing" creatures, he is not "watering down the concept of creation" but rather is "trying to give it, by remote analogies, some sort of content. I know that to create is defined as 'to make out of nothing,' *ex nihilo* … *not* out of any pre-ex-

---

1. Lewis, 1970, p. 21.
2. Lewis, 1970, p. 21.
3. Lewis, C. S. 1964. *Letters to Malcolm: Chiefly on Prayer.* New York: Harcourt, Brace & World, p. 91.

isting material."[4] Lewis adds that when

> we think that even human work comes nearest to creation when the maker has "got it all out of his own head." ...This act, as it is for God, must always remain totally inconceivable to man. For we—never, in the ultimate sense, *make*. We only build. We always have materials to build from. All we can know about the act of creation must be derived from what we can gather about the relation of the creatures to their Creator.[5]

In other words, "Creation," as applied to human authorship, is a misleading term; humans can only rearrange the elements that God has created. Only God can create *ex nihilo* (out of nothing).

Evolution as Mental and Spiritual Growth

Lewis often used the term "evolution" to refer to mental or spiritual growth, which created problems when he used this term in other contexts. For example, with "shrewd insight into the modern mind" Lewis turned "to evolution in order to explain what Christianity is all about. The 'new men' are presented as the next stage in evolution—or rather, as the stage that part of humanity has already reached."[6]

Another example is in his inaugural lecture at Cambridge given in 1954. The theme of this lecture was "actually illustrated in his earlier books" including his long essay, *The Abolition of Man*, and his science-fiction story *That Hideous Strength*.[7] In his lecture, Lewis proposed "that the largest shift in human culture was not, as usually claimed, the shift from Medieval to Renaissance but rather that which occurred early in the nineteenth century and introduced the post-Christian era." He also, in this novel foresaw the corruption of

---

4. Lewis, 1964a, pp. 72-73.
5. Lewis, 1964a, pp. 72-73.
6. Walsh, 1949, p. 28.
7. Duriez, 2013, p. 208.

the university, the rise of the expert class, and the marriage of science and politics.[8] No better example exists then science united to suppress critism of Darwinism by using "experts" and the courts to suppress objective evaluation of evolution in government schools. This is true in spite of the overwhelming scientific evidence in the peer reviewed literature against this worldview.[9] Lewis next

> expressed the surprising idea that Christians and ancient pagans had much more in common than either has with the post-Christian culture of today, and he enumerated four main areas in which the change could be observed: politics, the arts, religion, and "the birth of machines."[10]

This shift, Lewis concluded, facilitated the nineteenth-century belief in what he called spontaneous progress that "in turn owes something to Darwin's theory of evolution and perhaps to the 'myth of universal evolutionism,' which is different from Darwin and also pre-dates him."[11] When discussing "the Darwinian hypothesis ... he [Lewis] always makes a clear distinction between men and animals."[12] Conversely, he defines "universal evolutionism"

---

8. Cheaney, Janie. Mind Over Matter. CS Lewis Foresaw the corruption of the University. *World.* April 9, 2022.

9. Bergman, Jerry. 2022. *The Three Pillars of Evolution Demolished. Why Darwin was Wrong.* WestBow Division of Thomas Nelson and Zondervan and *The Minor Pillars of Evolution Demolished. Why Darwin was Wrong.* WestBow Division of Thomas Nelson and Zondervan.

10. Lewis, 1962. *They Asked for a Paper: Papers and Addresses.* London: Geoffrey Bles.; quoted in Kilby, Clyde S. 1964. *The Christian World of C. S. Lewis.* Grand Rapids, MI: Eerdmans, p. 174.

11. Kilby, 1964, p. 174.

12. Kilby, 1964, p. 174.

as "the belief that the very formula of universal process is from imperfect to perfect, from small beginnings to great endings, from the rudimentary to the elaborate: the belief which makes people find it natural to think that morality springs from savage taboos, adult sentiment from infantile sexual maladjustments, thought from instinct, mind from matter, organic from inorganic, cosmos from chaos." Though he believes this to be the very image of contemporary thought, he regards it as "immensely unplausible, because it makes the general course of nature so very unlike those parts of nature we can observe." We would do better, he says, to emphasize less that an adult human being came from an embryo that the embryo was produced by two adult human beings.[13]

From this description, it is transparently clear that the worldview Lewis opposes is orthodox Darwinism. The Christian view is

> precisely that the Next Step has already appeared. And it is really new. It isn't a change from brainy men to brainier men: it is a change that goes off in a totally different direction—a change from being creatures of God to being sons of God. The first instance appeared in Palestine two thousand years ago.[14]

The final pages of Lewis' book emphasized the price of becoming a Christian: "But there must be a real giving up of the self. You must throw it away 'blindly' so to speak. Christ will in fact give you a real personality."[15] Throughout his career, Lewis acknowledged the importance of humanity, and nowhere "is Lewis more proactive of the dignity of human nature than in defending it against the overblown use of science."[16]

---

13. Kilby, 1964, pp. 174-175.

14. Walsh, 1949, p. 28.

15. Walsh, 1949, p. 28.

16. Peterson, 2020, p. 135.

And as Chad Walsh recalled, Lewis observed that "the social sciences have a greater *de*humanizing potential than [even] the natural sciences."[17]

---

17. Peterson, 2020, p. 135.

Captain Bernard Acworth, president of the British *Evolution Protest Movement*. He was a prolific author of books on the living world. Creator: National Portrait Gallery London Credit: National Portrait Gallery London.

21

Statements Indicating Lewis Was a Theistic Evolutionist

SOME THEISTIC EVOLUTIONISTS argue that in certain of his early writings, such as *Mere Christianity*, Lewis appeared to accept, at least in part, some evolutionary ideas. As he researched and pondered on the subject, though, his writings eventually reflected a vivid opposition to the "Great Myth" of evolutionary naturalism. As Professors Ferngren and Numbers conclude, with study and reflection, "Lewis grew increasingly uncomfortable with the claims being made for organic evolution."[1] As noted, Numbers added that, privately, Lewis found the "arguments against evolution increasingly compelling—and the pretensions of many biologists repellant."[2]

Numbers then summarized the conversations Lewis had with Bernard Acworth (1885-1963), a "decorated World War I submariner and a pioneer in the development of sonar," which has revolutionized med-

---

1. Ferngren and Numbers, 1996, p. 28.
2. Numbers, 2006, p. 153.

icine with the invention of the sonogram. Acworth

> founded Britain's Evolution Protest Movement and published books criticizing evolution. It is not known when Acworth and C. S. Lewis first met, but the earliest of the ten surviving letters from Lewis to Acworth show that a warm friendship already existed in 1944, with Acworth sometimes staying with Lewis when he was in Oxford.[3]

Nonetheless, Lewis was reticent to express his conclusions about evolution publicly. A major reason Lewis shied away from openly endorsing the anti-evolution position of Acworth was because he [Lewis]

> feared that among his growing band of disciples some might take umbrage at his association with anti-Darwinists. [explaining:] 'When a man has become a popular Apologist ... he must watch his step. Everyone is on the lookout for things that might discredit him' Privately, however, he found Acworth's arguments against evolution increasingly compelling—and the pretensions of many biologists repellant.[4]

Much later, in 1951, Lewis openly

> confessed that his belief in the unimportance of evolution was shaken while reading one of his friends' manuscripts. 'I wish I were younger,' he confessed to Acworth. 'What inclines me now to think that you may be right in regarding it [evolution] as *the* central lie in the web of falsehood that now governs our lives is not so much your arguments against it as the fanatical and twisted attitudes of its defenders"[5]

And "Lewis observed there is a power behind the Great Myth (evolution as a philosophical position). All other views are subsidiary to the

---

3. Schultz, Jeffrey D. and John G. West. 1998. *Encyclopedia of Religion in American Politics.* Westport, CT: Greenwood Press, p. 69.

4. Numbers, 2006, p. 175.

5. Numbers, 2006, p. 175.

Great Myth... it serves as an eliminative enterprise."[6] In other words, other competing views, such as creationism, are forced from discussion, squelched and then suppressed by evolution.

In a review of *C. S. Lewis' Dangerous Idea,* Baylor University Professor of Philosophy C. Stephen Evans wrote that

> Darwinists try to show through science that our world and its inhabitants can be fully explained as the product of a mindless, purposeless system of physics and chemistry. But as Victor Reppert explains ... Lewis demonstrated that the Darwinian argument was circular: if such materialism or naturalism were true, then scientific reasoning itself could not be trusted. Reppert [demonstrates]... that—contrary to the dismissals of hasty critics—the basic thrust of Lewis' argument from reason can bear up under the weight of the most serious philosophical attacks.[7]

Evans concluded that Reppert carefully documented the fact that

> Lewis' most tenacious critics don't even attempt to refute his arguments [against Darwinism]—instead, they simply resort to base *ad hominem* attacks aiming squarely at modern-day intellectuals who sneer at Lewis' arguments because he wasn't a credentialed philosopher... Reppert demonstrates that Lewis' powerful philosophical instincts perhaps ought to place him among those other thinkers who, by contemporary standards, were also amateurs: Socrates, Plato, Aristotle, Aquinas, Descartes, Spinoza, Locke and Hume.[8]

Ironically, during his atheist years, "Lewis learned much of his skepticism from David Hume's famous *Essay on Miracles.*"[9] Thus, he

---

6. Loomis and Rodriguez, 2009, p. 148.
7. Evans, C. Stephen. 2004. "A Body Blow to Darwinist Materialism, Courtesy of the Great C. S. Lewis." Review of *C. S. Lewis' Dangerous Idea.* The Book Service.
8. Evans, 2004.
9. Hooper, Walter, 1996, p. 343.

knew an important part of his task as a Christian apologist was not only to respond to the problem of Darwinism and scientism, but also the problem of miracles.

In an effort to learn about the other side, while still an atheist, in 1926 Lewis questioned "the hardest boiled atheist he knew. He was shattered" when the atheist responded, not in ways to confirm Lewis' atheism as he had wanted then, but to confirm his doubts about atheism.[10] After carefully looking at both sides, as is well known, Lewis concluded that atheism was *not* supported by the facts of history, logic or science. He realized that, although "Darwin shocked the Victorians" it too "was *not* supported by the facts of history, logic or science."[11]

Authors Burson and Walls wrote that the great apologist Francis Schaeffer wrote an entire chapter on the creation-evolution conflict adding, nonetheless he "left the door open a crack to the possibility of theistic evolution – the position Lewis endorsed… though Schaeffer finds Lewis' hypothesis [about Adam and sin, discussed later] lacking biblical support, he [Schaffer] nonetheless presents it as a possibility."[12] If Lewis also presented theistic evolution as a possibility, which he did not, this does not mean he was a theistic evolutionist, only that he did not dogmatically outright reject this possibly. As we will show, as Lewis grew older, he moved solidly farther away from this worldview.

Although a few statements that Lewis made may appear to indicate that he was a theistic evolutionist, when examined in context, and in view of the stage of his intellectual and spiritual growth, these statements do not openly support this conclusion. In what claims to be a "comprehensive study" of Lewis' views on Intelligent Design and evolution, Theology Professor Michael L. Peterson quoted what Lew-

---

10. Green and Hooper, 1974, p. 102.
11. Hooper. Walter, 1996. p. 486.
12. Burson and Walls, 1998, pp. 137-138.

is wrote in *Mere Christianity*[13] as saying "Perhaps a modern man can understand the Christian idea [of transformation] best if he takes it in connection with Evolution. Everyone now knows ... that man has evolved from lower types of life."[14]

However, Professor Wile notes: "This quote makes it sound like Lewis firmly believed that man evolved from lower life forms such as apes. However, that's not what Lewis wrote."[15] The entire quote, with the sections that Peterson lifted in bold, is as follows:

> **Perhaps a modern man can understand the Christian idea best if he takes it in connection with Evolution. Everyone now knows** about Evolution (though, of course, some educated people disbelieve it): everyone has been told **that man has evolved from lower types of life**.[16]

As Wile correctly observed, this

> passage says something quite different from what Peterson wants you to believe. It doesn't at all imply that Lewis believes man evolved from lower types of life. It doesn't even imply that Lewis thinks everyone else does. In fact, he specifically says that he knows there are some educated people who don't believe in it. So Lewis was making a much more tentative statement about man evolving from lower forms of life than what Peterson wants you to think he was making. It is unfortunate that Peterson quotes Lewis out of context, simply to make it look like Lewis agrees with him.[17]

The passage in its entirety clearly shows that, far from asserting

---

13. Lewis, C. S.. 1960. *Mere Christianity.* New York: Collier Publishing, p. 154.
14. Lewis, 2001a, p. 218.
15. Wile, Jay. 2011. "Thoughts from a Scientist who is a Christian (Not a Christian Scientist)." *Everyone Wants a Piece of C. S. Lewis.* Friday, July 22.
16. Wile, 2011, p. 18.
17. Wile, 2011, p. 1.

"evolution" is something that "everyone now knows," Lewis was stating only that "everyone now knows *about* Evolution," and "everyone has been *told*" about it from school teachers, the media, friends and other sources. He here was describing the popular view. He was not claiming that human evolution is true, or even that he believed it was true. Lest someone misunderstand that Lewis was endorsing the view of evolution that "everyone has been told", Lewis added the caveat, "of course, some educated people disbelieve it."[18]

Even if Professor Peterson's interpretation of Lewis' view on evolution was valid, it must be remembered that the quote was taken from *Mere Christianity* not long after Lewis converted from atheism, and he became more hostile towards Darwinism as he learned more about the harm it had caused. Lewis had major problems with common descent, as he studied and thought about the implications of the doctrine in more detail. To lift one section from his lifelong writings and claim that "clearly... Lewis accepts the Darwinian concept of 'common descent,'" as Peterson did, would be like concluding that, based on his early beliefs, Lewis clearly was an atheist.[19] In *Christian Reflections*, Lewis writes it is his view that to

> the biologist, Evolution is a hypothesis. It covers more of the facts than any other hypothesis at present on the market, and is therefore to be accepted unless, or until, some new supposal can be shown to cover still more facts with even fewer assumptions. At least, that is what I think most biologists would say.[20]

Of course, this view is an outsider's evaluation of what Lewis *assumed* biologists believed, not what he [Lewis] believed. Lewis also qualified his statement above, stating he had major doubts about com-

---

18. West, 2012, p. 112.
19. Peterson, 2010, p. 260.
20. Lewis, 1967, p. 83.

mon descent because

> Darwinism gives no support to the belief that natural selection, working upon chance variations, has a general tendency to produce improvement. The illusion that it has comes from confining our attention to a few species which have (by some possibly arbitrary standard of our own) changed for the better. Thus the horse has improved in the sense that *protohippos* [*Protohippus*] would be less useful to us than his modern descendant ... But a great many of the changes produced by evolution are not improvements by any conceivable standard... There is no general law of progress in biological history.[21]

Lewis accepted that microevolutionary change could occur by evolution, but not evolutionary "progress in biological history" which is what Darwinism is all about, such as eyespots evolving into vertebrate eyes from many hundreds of mutations. As Lewis wrote, "Evolution is not only not a doctrine of *moral* improvements, but of biological changes, some improvements, some deteriorations."[22]

The Penelope Letter

In researching Lewis, I commonly came across irresponsible claims, such as those made using the following quote in a letter Lewis wrote to Sister Penelope on October, 1, 1952:

> "I, like you, had pictured Adam as being, physically, the son of two anthropoids, on whom, after birth, God worked the miracle which made him Man." Lewis is here *explicitly denying spontaneous creation and affirming that Adam was born of non-human primitive ancestors.*[23]

---

21. Lewis, 1960, p. 103.

22. Lewis, 1960a, p. 218; Lewis, C. S. 1966. *Letters of C. S. Lewis.* Grand Rapids, MI: Eerdmans, pp. 392-393.

23. https:// amazon.com/product-reviews/1532607733/ref=acr_dp_hist_1?ie=UTF8&filterByStar=one_star&reviewerType=all_reviews#reviews-filter-bar. Em-

Interpreting the Letter

This exchange was not about the evolution of mankind from apes, or even about evolution. The letter Lewis answered was triggered by the letter which Penelope, an Anglican nun, wrote to Lewis in response to an article she read about the discovery of footprint evidence for the Abominable Snowman called Yeti, meaning *the wild man of the snows*.[24] This giant manlike creature has been part of Western folklore since about 1921.[25]

Lewis wrote in response to her question, explaining that "I, like you, had pictured Adam as being, physically, the son of two anthropoids, on whom, after birth, God worked the miracle which made him Man."[26] Contrary to the common interpretation that this refers to human evolution and teaches God used an ape body on the way to evolving into a human to create humans, and that God made a man by infusing a soul into this ape body, the letter shows something very different. As Lewis writes, the concern relates to

> the old problem Who was Cain's wife? If we follow the scripture it wd [would] seem that she must have been no daughter of Adam's… the solution was regeneration in each one of us wd [would] be an instance too… the call of Abraham wd [would] be a smaller instance of the same sort of thing," and regeneration in each one of us wd [would] be an instance too, tho' of the same process, tho' not a smaller one."[27]

Lewis rejected the other possibility that some have suggested, phasis added.

---

24. *The [London] Times*. 6 December 1951, p. 5. [Headline of article? Not listed in References section.]
25. Pranavananda, Swami. 1957. "The Abominable Snowman." *Journal of the Bombay Natural History Society* **54**(1-2):179-181. (Article titled "Albino Elephants")
26. Lewis, 2007, p 157.
27. Lewis, 2007, p. 157.

namely that "Adam was a hermaphrodite." White's comments on this letter remind us that "Lewis considered man to be participating even yet in the process of evolution. Or better still, he said, man continues 'in the process of being created.'"[28] The later reading may be a far more accurate summary of Lewis' thoughts. Lewis was not endorsing human evolution from an ape ancestor but something very different, namely about man's *maturity* which he expressed as man "in the process of being created."

This concern of where Cain got his wife may seem naive today because the answer is obvious: he married his sister or a niece, likely a woman he had barely known since he evidently lived some distance away from her and, genetically, he and his sister/niece were close to perfect without many, if any, mutations. It would require many generations to accumulate the mutational load that causes the genetic problems we recognize today when close relatives marry.

According to Jewish historian Flavius Josephus (born AD 37 or 38; died AD 100), when Adam was 230 years old, his wife had yet another child to add to his many children.[29] Ancient Jewish tradition teaches that Adam and Eve had at least 33 sons and 23 daughters. Adam's son Enoch alone had 77 children by his two wives, Silla and Ada.[30] Thus, the early population increased rapidly and no doubt Cain did not grow up with his sister. Lewis noted that the "True Men descended from Seth," the third son of Adam and Eve. His brothers Cain and Abel were somehow not "true men" in the same way that Seth was, but they were not apes or pre-humans. Thus, Lewis' "true men" reference does not relate to evolution, but likely their standing with God.

---

28. White, 1969, p. 101.

29. Whiston, William (Translator). 1987. *The Works of Flavius Josephus*. Peabody, MA: Hendrickson Publishers, p. 32.

30. Whiston, 1987, p. 31.

That Lewis was not explicitly denying spontaneous creation nor affirming that Adam was born of non-human primitive ancestors is obvious from his other writings. White, in his Ph.D. dissertation, wrote: "People sometimes wonder what the next step in evolution might be." White explains a similar thought as Lewis and, putting Lewis' words into modern language, would be equal to "the human race is right now embarking on its next major phase of maturity, as men change from being creatures of God's to becoming sons of God… to forget their old natural selves and to move into this entirely new realm. Here they can begin to become what they were designed by their Creator to be." [31]

---

31. White, 1969, pp. 140-141.

Molecules to Man Evolution is taught in many major biology textbooks, such as the one pictured above published by Cambridge University Press.

22

Lewis and the Genesis Fall

In Lewis' mind, the Genesis Fall was a critical Christian doctrine that he discussed at length, both in his fiction and nonfiction works. In a study of the Adam-Eve doctrine, Collins wrote that Lewis' position on a literal Adam and Eve in his book *The Problem of Pain* is vague. In this book, Lewis calls it an "imaginative exercise," but "in his other books, he [Lewis] keeps to a particular Adam and Eve, as he has great respect for .... the story in Genesis."[1] Lewis admitted that "I had also an evolutionary background which led me to think of early men, and therefore *a fortiori* [with convincing force] of the first men, as savages. The beauty I expected in Adam and Eve was that of the primitive, the unsophisticated.... I wanted an Adam and Eve whom I could patronize: and ... Milton made it clear that I was not to be allowed to do anything of the sort."[2]

Lewis makes it clear in his writings that his early view was clarified. For example, he wrote, as Adam and Eve were created *ex-nihilo* and did

---

1. Collins, 2011, p. 130.
2. Lewis, 1942, p. 116.

not evolve, they "were never young, never immature or underdeveloped" because they would have been created as mature adults.[3] Then he wrote, "Man has fallen from the state of innocence in which he was created. Therefore, I disbelieve any theory that contradicts this."[4] This same conclusion often was expressed in his science-fiction stories. For example, in the sequel to the *Silent Planet,* humans were still innocent, and thus Lewis states they were at the Adam and Eve stage.[5] Lewis' imaginative exercise has "moved us away from the Biblical Story," but, nonetheless, he

> preserves the *historical character of the fall*: that is, it is an event—or cluster of events—that actually took place, and changed human life forever. This certainly sets his view apart from all views that see sin as the result of something "timeless and eternal" and thus non-historical (as in Karl Barth), or as something inherent in God's creation (as in many modern theologians).[6]

Collins also noted that the major difficulty he had

> with Lewis' view lies in his clause, "We do not know *how many* of these creatures God made." He is not asserting that there *must* have been more than Adam and Eve; he is declaring the question immaterial to the discussion. If, however, we take our cue from Lewis' own mention of solidarity and "in Adam," ... then we have a way of pushing Lewis' scenario into a more acceptable direction. That is we should make it more like ... Adam as the chieftain and Eve as his queen.[7]

---

3. Lewis, 1942, p. 116.
4. Lewis, 2004b, p. 842. Letter dated September 23, 1944.
5. Dickerson, Matthew T. and David O'Hara. 2009. *Narnia and the Fields of Arbol: The Environmental Vision of C. S. Lewis.* Lexington, KY: University Press of Kentucky, p. 186.
6. Collins, 2011, p. 130. Emphasis added.
7. Collins, 2011, p. 130.

Lewis himself also wrote in *The Problem of Pain* that even the small details of the Fall may well be literal: "For all I can see, it might have concerned the literal eating of a fruit."[8]

Man Physically Descended from Animals

Another place where Lewis discusses the Genesis Fall is also in *The Problem of Pain*. This book was the first of a series of popular works that Lewis wrote on Christian doctrine. As the book was authored in 1940, not long after Lewis converted to Christianity from agnosticism, some claim he might have been somewhat ambivalent then on the subject of evolution. For this reason, Lewis in this book considered the problem of suffering from a purely theoretical standpoint, which he implied was valid *even if* theistic evolution were true.

When arguing specifically for the orthodox Genesis Fall-of-man doctrine, Lewis made it clear that the account requires accepting a sinless, perfect Adam, or at least a man that was a fully human species, and a fall from grace. Why then, did Lewis write, "if by claiming that man rose from brutality you mean simply that man is physically descended from animals, I have no objection."[9]

Lewis did *not* say he agreed with this view, but that he had "no objection," if people wanted to believe this view, possibly because Lewis here was arguing that the doctrine of the Genesis Fall could be defended, *even* if one tried to argue "that man is physically descended from animals." Likely, one reason is because even this view still allows for the acceptance of the Fall.

Considering the context, it seems clear to me that Lewis was not referring to the idea that humans evolved from ape ancestors, but physically descended from primitive men that have slowly risen *from brutality and savagery,* meaning they have *behaved* like animals, not that

---

8. Kilby, 1964, p. 154; Lewis, 1996d, pp. 76-77.
9. Lewis, 1996d, p. 67.

they were some kind of monkey that evolved into humans. This is clear from what Lewis wrote:

> If by saying that man rose from brutality you mean simply that man is physically descended from [men that behaved like] animals, I have no objection. But it does *not follow that the further back you go the more brutal—in the sense of wicked or wretched—you will find man to be.*[10]

This view is supported by his previous statement that the creation doctrine specifically teaches that

> Man as God made him, was completely good and completely happy, but that he disobeyed God and became what we now see. Many people think that this proposition has been proved false by modern science. 'We now know,' it is said, "that so far from fallen out of primeval state of virtue and happiness, men have slowly risen from brutality and savagery." There seems to me to be a complete confusion here. *Brute* and *savage* both belong to that unfortunate class of words which are sometimes used rhetorically, as terms of reproach, and sometimes scientifically, as terms of description, and the pseudo-scientific arguments against the Fall depends on a confusion between the usages.[11]

These pseudo-scientific arguments against the Fall are what this section of *The Problem of Pain* is discussing, not human evolution from an ape ancestor. That the word *animals* refers to *behavior* is obvious from the text, as well as Lewis' comments about what he means by "saying that man rose from brutality you mean simply that man is physically descended from [men that behaved like] animals, I have no objection," as is shown below:

> No animal has moral virtue; but it is not true that all *animal behavior* is of the kind one should call "wicked" if it were practiced by men. On the

---

10. Lewis, 1996d, p. 67. Emphasis added.
11. Lewis, 1996d, p. 67.

contrary, not all animals treat other creatures of their own species as badly as men treat men. Not all are as gluttonous or lecherous as we, and no animal is ambitious. Similarly if you say all men were "savages, meaning by this that their artifacts were few and clumsy like those of modern savages, you may well be right, but if you mean they were "savage" in the sense of being lewd, ferocious, cruel, and treacherous, you will be beyond your evidence, and that for two reasons. In the first place you cannot argue from the artifacts of the earlier men that they were in all respects like the contemporary people who make similar artifacts... Prehistoric man, because he is prehistoric, is known to us only by the material things he made.[12]

Note that in his writings, Lewis used the term "savages" and "savage mind," as well as "primitive" to refer to "savage ideas," such as those of the pre-Copernican world as well as modern "savages" such as young, immature school children.[13] I know of no case where Lewis clearly used the terms to refer to pre-human evolutionary ancestors.

The orthodox interpretation of man's Fall into sin, as revealed in the Scriptures, involves not brute prehuman bodies evolving into modern bodies, but *descending* into behavior that is cruel, treacherous, ferocious, or evil. The existence of pre-industrial societies in prehistory that evolutionists believe to be primitive does not negate the Fall, as Lewis takes pains to document.[14] Lewis argues that these so-called primitive societies that evolutionists believe existed are a myth. In contrast to the teachings by evolutionists, Lewis writes that the

> modern estimate of primitive man is based upon that idolatry of artifacts which is a great corporate sin of our own civilization. We forget that our

---

12. Lewis, 1996d, pp. 66-67.
13. For example see C. S. Lewis. 2013. *Studies in Medieval and Renaissance Literature*. Cambridge: Cambridge University Press, pp. 11, 14, 42, 43, 61, 126.
14. Lewis, 1996d, p. 68.

prehistoric ancestors made all of the useful discoveries, except that of chloroform, which have ever been made. To them we owe language, the family, clothing, the use of fire, the domestication of animals, the wheel, the ship, poetry and agriculture. Science, then, has nothing to say against the doctrine of the Fall.[15]

Roberts noted that, in Lewis' novels, "the literal reality of Lewis' mythical figures (the Savior/Fisher King, the Devil, Adam and Eve) is only possible via Lewis' conservative Christian faith."[16] In a letter to a Miss Breckenridge, dated August 1, 1949, Lewis wrote that there

> is *no* relation of any importance between the Fall and Evolution. The doctrine of Evolution is that organisms have changed, sometimes for what we call (biologically) the better ... quite often for what we call (biologically) the worse ... The doctrine of the Fall is that at one particular point one species, Man, [specifically Adam] tumbled down a moral cliff. There is neither opposition nor support between the two doctrines ... Evolution is not only not a doctrine of moral improvements, but of biological changes, some improvements, some deteriorations.[17]

Lewis is here clearly referring to change in time, not molecules-to-man evolution. He writes that he defines evolution as "organisms have changed, sometimes for what we call (biologically) the better ... quite often for what we call (biologically) the worse," a definition any breeder knows is an accurate description of the variation within the Genesis kinds. Lewis' definition of evolution is hardly the orthodox macroevolution view which postulates an evolution from simple organic compounds to modern humans.

---

15. Lewis, 1996d, pp. 68-69.

16. Roberts, Adam. 1998. *Silk and Potatoes: Contemporary Arthurian Fantasy*. Atlanta, GA: Rodopi, p. 23.

17 Lewis, 2004b, p. 962. Emphasis in original. The letter to Ms. Breckenridge is dated August 1, 1949. Reprinted in *Collected Letters*, Volume 3, 2006, p. 962.]

In another article, even though he made his objections to Darwinism perfectly clear, Lewis wrote "*for the purpose of this article* I am assuming that Darwinian biology is correct", again attempting to show that his particular argument was still valid, even if Darwinian evolution was true. White did the same thing in his thesis about Lewis, writing, "Whether conscious of the reason or not, some people feel the idea of the Second Coming [of Christ] is out of character with the dominant emphasis today upon evolution and gradual development. Assuming Darwinian biology to be correct, there is, of course, no gradual law of progress in biological history."[18] Professor White was not endorsing evolution but, even assuming biological evolution were true, was making a statement about the implications of the belief which would hold true whether evolution were true of not.

Lewis uses the same reasoning in the passage that Peterson "cherry-picked" to prove that he [Lewis] was a theistic evolutionist, or even a Darwinist. The passage is as follows, with the parts that Peterson selected in bold.[19] This passage was also first published in 1940, not long after Lewis' conversion from atheism and therefore represents his early thoughts about Christianity and creation:

> **For long centuries God perfected the animal form which was to become the vehicle of humanity and the image of Himself.** He [God] gave it hands whose thumb could be applied to each of the fingers, and jaws and teeth and throat capable of articulation, and a brain sufficiently complex to execute all the material motions whereby rational thought is incarnated. **The creature may have existed for ages in this state before it became man**: it may even have been clever enough to make things which a modern archaeologist would accept as proof of its humanity. But it was only an animal because all its physical and psychical processes were directed to purely material and natural ends. Then, **in the fullness**

---

18. White, 1969, p. 192.
19. Peterson, 2010, p. 260.

of time, God caused to descend upon this organism, both on its psychology and physiology, **a new kind of consciousness which could say "I" and "me"**, which could look upon itself as an object, **which knew God**, which **could make judgments of truth, beauty, and goodness**, and which was so far above time that it could perceive time flowing past. This new consciousness ruled and illuminated the whole organism flooding every part of it with light… Man was then all consciousness.[20]

This passage is a hypothetical story that refers, not to human evolution from apes but to human mental and spiritual growth. This interpretation is consistent with what Lewis wrote in his "The Funeral of a Great Myth" essay,[21] *where* Lewis makes it clear that he is willing to accept variation within the Genesis kinds or microevolution, but he is not willing to accept macroevolution, such as the evolution of all higher forms of life from single-celled animals.[22] Lewis wrote:

> for the scientist Evolution is purely a biological theorem. It takes over organic life on this planet as a going concern and tries to explain certain changes within that field. It makes no cosmic statements, no metaphysical statements, no eschatological statements [as does Darwinism].[23]

He adds that

> we now have minds we can trust, granted that organic life came to exist, it tries to explain, say, how a species that once had wings came to lose them. It explains this by the negative effect of environment operating on small variations. It does not in itself explain the origin of organic life, nor

---

20. Lewis, 1996, p. 72. Bold was added to show the section Peterson quoted.
21. Lewis, 1967, p. 85.
22. See Corwin, 2016, p. 7, where he makes it clear that Lewis was referring to evolution.
23. One of many sources of the essay *Funeral of a Great Myth* is in the book *Christian Reflections*.

of the variations, nor does it discuss the origin and validity of reason. **It may well tell you how the brain, through which reason now operates, arose**, but that is a different matter.[24]

Peterson uses this quote to claim that "Clearly, Lewis accepts the Darwinian concept of 'common descent with modification.'"[25] However, the full quote and the context of the chapter, shed a very different light on Lewis' words than Peterson implied. Lewis *prefaced* the above quote by stressing that "What exactly happened when Man fell, we do not know, but *if it is legitimate to guess*, I offer the following picture—a 'myth' in the Socratic sense, a not unlikely tale."[26] One interpretation of a myth in the Socratic sense is a "story that *may have been* historical fact."[27]

The word myth, as used in this context, was explained by Edmund Fuller, who opined "that Lewis' science-fiction trilogy represented a fresh telling of the Christian myth of the Fall of man. So used, myth does not mean fiction, of course, but truth through symbolism."[28] Fuller added that Lewis

> dramatizes and clarifies for our age the Christian teaching about man's peculiar dilemma in the order of Creation. The tragic fact of man's condition is that he is other than he was intended to be; the deep springs of his will have been subverted—he cannot do consistently the good that he would, but does instead the evil that he would not do.[29]

---

24. Lewis, 1967, p. 107.
25. Peterson, 2010, p. 260.
26. Lewis, C. S. 1947. *The Problem of Pain*. New York: MacMillan, p. 76.
27. Collins, 2011, p. 128.
28. White, 1969, p. 128.
29. Fuller, Edmund. 1962. *Books with Men Behind Them*. New York: Random House, p. 155.

Schmerl spoke of this science-fiction trilogy as "an imaginative extension of Christian mythology, centering around a new battle between God and his angels and Satan and his devils, fought on three planets."[30]

Nonetheless, Lewis was concerned that science, in order to replace God, must create its own dogma, and "Lewis knew of the inconsistency of evolutionary theology and ethics, pointing out that ... ideas of aesthetics, morality, and humanity cannot logically be derived from a random chance universe."[31] The source of these observations shows that Lewis fully recognized the implications of evolution, writing:

> Reason has Evolved 'out of instinct, virtue out of complexes, poetry out of erotic howls and grunts, civilization out of savagery, the organic out of inorganic, the solar system out of some sidereal soup or traffic block. Conversely, reason, virtue, and civilization as we now know them are only crude or embryonic beginnings of far better things—perhaps Deity itself—in the remote future. For in the Myth, "Evolution (as the Myth understands it) is the formula of *all* existence.[32]

Lewis then continues to show the absolute absurdity of Darwinian evolution, writing:

> To exist means to be moving from the status of 'almost zero' to the status of 'almost infinity.' To those brought up on the Myth nothing seems more normal, more natural, more plausible, than that chaos should turn into order, death into life, ignorance into knowledge. And with this we reach the full-blown Myth. It is one of the most moving and satisfying world dramas which has ever been imagined.[33]

---

30. Schmerl, Rudolf B. 1960. "Reasons Dream: Anti-Totalitarian Themes and Techniques of Fantasy." Ann Arbor, MI: University of Michigan, p. 60. [Schmerl died on 5/7/2020.]

31. Corwin, 2016, p. 47.

32. Loomis and Rodriguez, 2009, p. 77.

33. Loomis and Rodriguez, 2009, p. 77.

Lewis then shows the harm Darwinism has done to society, writing while mocking Darwinism: "Eugenics have made certain that only demi-gods will now be born: psycho-analysis that none of them shall lose or smirch his divinity: economics that they shall have to hand all that demi-gods require. Man has ascended his throne. Man has become God. All is a-blasé of glory."[34]

No overview of Lewis and evolution would be complete without reference to Lewis' portraits of lost and evil men in his science-fiction trilogy, *Out of the Silent Planet, Perelandra,* and *That Hideous Strength*. From the cold contempt and unscrupulous exploitation of feeble-minded Harry and Weston in *Planet*, to Filostrato's experiment with a guillotined man's head and Wither's trance-like senility in a demon-made void in *Strength*, exists a wealth of prophetic realism about the end result of the emergent evolutionist worldview for its practitioners' victims.

Most horrible of them all is Weston, "a convinced believer in emergent evolution… the goal towards which the whole cosmic process is moving. Call it a great, inscrutable Force, pouring up into us from the dark bases of being… your Devil and your God are both pictures of the same Force." From these excerpts it is clear that Lewis was unalterably opposed to emergent evolutionism and would likely have welcomed the rise of creation science that developed in our own generation.

In another letter, dated December 29, 1958, Lewis described his novel *Perelandra* as working out of the 'supposition' that what happened to Adam and Eve on Earth could happen to another first couple elsewhere: "Suppose, even now, in some other planet there were a first couple undergoing the same [temptation] that Adam and Eve underwent here [on Earth], but successfully."[35]

---

34. Lewis, 1967, p. 88.

35. Lewis, 2007, p. 1004.

Lewis even viewed the vast astronomical distance in the universe as "God's quarantine precautions" to "prevent the spiritual infection of a Fallen species from spreading" to other planets.[36]

Perelandra is the second volume of Lewis' science-fiction trilogy. What is particularly significant in this volume is the Garden of Eden example on the Planet Perelandra (Venus), which was a paradisiacal ocean-world that is an analogue of the Garden-of-Eden account. According to Fuller, "The tempting of the first woman of Perelandra entails an extraordinarily intricate, far-reaching debate at the deepest level of moral theology."[37]

Analysis of the "Not Unlikely Tale"

"For long centuries God perfected the animal form" does not refer to perfecting or evolving an ape body to become a human, but rather God perfecting the animal form by adding the spiritual-mind component to the animal body. Lewis called humans who lacked spiritual development 'animals', writing "it was only an animal because all its physical and psychical processes were *directed to purely material and natural ends*" but it was to become the vehicle of humanity and the image of Himself.

Although he had hands, fingers, jaws and "teeth capable of articulation, and a brain sufficiently complex to execute all the material motions whereby rational thought is incarnated" it did not have a "consciousness which could say "I" and "me," which could look upon itself as an object, which knew God" and "could make judgments of truth, beauty, and goodness." The man-beast comparison is also found in the Bible – specifically Ecclesiastes 3:18-21 and 2 Peter 2:12.

Only when it had this mind ability did it become man. And this "new consciousness ruled and illuminated the whole organism flood-

---

36. Deasy, 1958, p. 422.

37. White, 1969, p. 128.

ing every part of it with light... Man was then all consciousness." From this analysis it is clear that this account refers, not to the evolution from some ape ancestor, but, as Lewis plainly stated, refers only to "what happened when Man fell."[38]

The rest of this chapter in *The Problem of Pain* strongly supports the view that Lewis personally believed in a literal Adam.[39] For example, Lewis wrote that the Holy Spirit would not have allowed the inclusion of the creation account of Adam and Eve and the Fall of Man in the Bible if they were not true historical events. Furthermore, this view would not have achieved "the assent of great [church] doctors unless it ... was true."[40] Lewis adds that the Adam and Eve account about "the magic apple [which]... brings together the trees of life and knowledge, contains a deeper and subtler truth than the version which makes the apple simply and solely a pledge of obedience."[41]

In the chapter titled "The Fall of Man" in *The Problem of Pain*, Lewis identifies Adam and Eve's original sin was that they somehow assumed that "they could become as gods—that they could cease directing their lives to their Creator" God, wanting to "take care for their own future, to plan for pleasure" but that is to live a lie because "our souls are not, in fact, our own. They wanted [to inhabit] some corner of the universe of which they could say to God, 'This is our business, not yours.' But there is no such corner. They wanted to be nouns, but they were, and eternally must be, mere adjectives."[42]

In short, from a review of numerous passages, it is safe to agree with Lewis scholar and biographer Professor Colin Duriez's conclusion that

---

38. Lewis, 1996d, p. 71.
39. Lewis, 1996d, pp. 63-85.
40. Lewis, 1996d, p. 66.
41. Lewis, 1996d, p. 66.
42. Lewis, 1996d, p. 75.

Lewis "believed in a real historical fall by disobedient mankind that affected the whole of nature."[43] One of many reasons Lewis opposed evolutionism was because "evolutionism was opposed to the doctrine of original sin."[44] Duriez noted that Lewis defined Adam and Eve as "the first humans with all people in every part of the world descending from them."[45]

This is one reason why in his novels, such as *The Chronicles of Narnia*, the throne may be occupied only by "the sons of Adam and the daughters of Eve."[46] Lewis also believed that the Fall affected the animals as well as humans.[47] In other words, Lewis often described the Fall in such a way that it required not only a literal Adam, often using the pronoun "he," but the origin of all creation (including all life on Earth) and the plants, must be understood in the literal sense.

Lewis' novel *Perelandra* also deals with the implications of evolution and Lewis' opposition to social Darwinism.[48] He rejected the "moral and philosophic implications that are popularly thought to follow from" evolution.[49] Lewis did not explain the evolutionary idea to convey his own beliefs, but to show the evil implications of the theory.

To achieve this goal, Lewis had to present it in all its implications, especially those of the Social Darwinists and the idea of "survival of the fittest" which, in the novel, resulted in two races, the race Lewis called *Elio* which developed from the aristocrats, and the *Morlocks*, a race

---

43. Duriez, 2013, p. 215.
44. Dickerson and O'Hara, 2009, p. 278.
45. Duriez, 2013, p. 15.
46. Duriez, 2013, p. 81.
47. Lewis, 2004b, p. 460.
48. Myers, Doris. 1994. *C. S. Lewis in Context*. Kent, OH: Kent State University Press, p. 47.
49. Myers, 1994, p. 56.

which evolved from the working class.⁵⁰ Lewis called Social Darwinism "regenerative science," writing that a decent person "would not do even to minerals and vegetables what modern science threatens to do to man himself."⁵¹ In short,

> Lewis' target is the modern evolutionary or "developmental" paradigm, but in this novel [*Perelandra*] the emphasis shifts from the materialist "struggle for existence" to Henri Bergson's more affirmative vitalist philosophy of creative (or emergent) evolution. Just as Martian civilization represents a transfiguration of the Darwinian view of the evolutionary process, the ever developing and open-ended character of the creation on Venus suggests that this new Eden is a sublimated version of creative evolution itself.⁵²

Conversely, in his novels, Lewis did not use the "struggle for existence" example or the "appropriation of Wells' evolutionary naturalism" to argue for evolution from a single-celled life-form to humans. Rather, he used these examples to show the reality of "red in tooth and claw," a reference to the sometimes violent natural selection in the world when predatory animals unsentimentally cover their teeth and claws with the blood of the prey they kill and devour.⁵³ His concern was the teachings of "Marx, Lyell, and Darwin whose teachings," Lewis concluded, caused us to "Deny the existence of free will."⁵⁴

---

50. Myers, 1994, p. 59.

51. Quoted in Myers, 1994, p. 84. From C. S. Lewis *The Abolition of Man* p. 89-90.

52. Schwartz, 2009, Fromhttps://www.oxfordscholarship.com/view/10.1093/acprof:oso/9780195374728.001.0001/acprof-9780195374728-chapter-3.

53. Schwartz, 2009, p. 14.

54. Myers, 1994, p. 33.

Helen Gardner Fellow of St Hilda's College from 1942 to 1966.

23

Evolution Cannot Explain the Origin of the Mind or Life

We argued earlier that consistent with Lewis' 'macroevolution-no/microevolution-yes' stance in his "The Funeral of a Great Myth" essay,[1] Lewis stated, paraphrasing that "evolution *may well tell you how the brain, through which reason now operates, arose* but not how the mind arose." This would argue that even if physical processes could explain the corporeal container of man's reasoning ability, it could never account for the origin of its incorporeal contents. After all, Lewis noted, all mammals have a brain, but not one mammal has a mind even close to humans. Lewis here may be referring, not to the evolution of the brain from simple one-celled life, but to modern mankind's developed, cultured, educated mind compared to those of men living in primitive societies, such as the Neanderthals, albeit a people group now accepted as fully human.[2]

---

1. Lewis, 1967, p. 85.
2. Bergman, et al, 2020.

Note that Lewis says evolution *cannot* explain either the origin of the mind or the origin of life. Thus, he is clearly not an evolutionist as commonly defined. When Lewis wrote most of his early books, he was still being careful not to openly challenge the evolutionary establishment. Thus, in his "The Funeral of a Great Myth" essay, he wrote that evolution "may well tell you *how the brain*, through which reason now operates, *arose*" but evolution does not *explain how the brain evolved.*

The fact is, Lewis wrote many seemingly contradictory statements that, because he is no longer with us, have to be interpreted in the context of his entire life's work. Even though one could lift a few quotes from Lewis' writing in an attempt to prove the evolutionary case, to be consistent with all of his writings related to this topic, the few places where Lewis *appears* to support macroevolution must be examined.

The Helen Gardner Event

Evidence for the view presented here includes an event documented by a person Lewis referred to as "A.N." based on an interview with his Oxford colleague, Professor Helen Gardner. Lewis biographer A.N. (Andrew Norman) Wilson wrote that, at a dinner party with friends, the conversation turned to the interesting question of who those present would most look forward to personally meeting after death. One person answered Shakespeare, another St. Paul. When Lewis was asked the same question, he answered, "I have no difficulty in deciding, I want to meet Adam." He went on to explain why in very much the same terms he outlined in *A Preface to 'Paradise Lost,'* where Lewis wrote that

> Adam was, from the first, a man in knowledge as well as in stature. He alone of all men 'has been in Eden, in the garden of God: he has walked up and down in the midst of the stones of fire.' He was endowed, says Athanasius, with 'a vision of God so far-reaching that he could contemplate the eternity of the Divine Essence and the cosmic operations of His Word.' He was 'a heavenly being,' according to St. Ambrose, who

breathed the aether, and was accustomed to converse with God 'face to face.' 'His mental powers,' says St. Augustine, 'surpassed those of the most brilliant philosopher as much as the speed of a bird surpasses that of a tortoise.'[3]

Wilson comments that Adam is not likely

> to converse with [Oxford Professor] Helen Gardner. ... she told Lewis, if there really were, historically, someone whom we could name as 'the first man,' he would be a Neanderthal ape-like figure, whose conversation she could not conceive of finding interesting. A stony silence fell on the dinner table. Then Lewis said gruffly, 'I see we have a Darwinian in our midst.' Helen Gardner was never invited [to dinner] again. Another Oxford woman with whom Lewis famously crossed swords at this period was Elizabeth Anscombe, the philosopher.[4]

This exchange also supports the view that Lewis believed in a literal Adam and could not have been a Darwinist.

Lewis writes in Chapter 16 of *A Preface to 'Paradise Lost'* that he had come to Milton's poem *Paradise Lost* associating the innocence of Adam with childishness. Lewis also confessed to yet another mental obstacle related to Adam: "I had also an evolutionary background which led me to think of early men, and therefore *a fortiori* of the first men, as savages." Consequently, due to this bias, what he

> expected in Adam and Eve was that of the primitive, the unsophisticated, I had hoped to be shown their inarticulate delight in a new world which they were spelling out letter by letter, to hear them prattle. Not to put too fine a point on it, I wanted an Adam and Eve whom I could patronize; and when Milton made it clear that I was not to be allowed to do anything of the sort, I was repelled.

---

3. Lewis, 1942, p 117. Also quoted in Wilson, A. N. 1990. *C. S. Lewis: A Biography*. New York: Norton, p. 210.

4. Wilson, 1990, p. 210.

Lewis adds that Milton gives a glimpse of what Adam would be like if he had never sinned, and thus never had fallen:

> He would still have been alive in Paradise, and to that "capital seat" all generations "from all the ends of the Earth" would have come periodically to do their homage (XI, 342). To you or to me, once in a lifetime perhaps, would have fallen the almost terrifying honor of coming at last, after long journeys and ritual preparations and slow ceremonial approaches, into the very presence of the great Father, Priest, and Emperor of the planet Tellus; a thing to be remembered all our lives. *No useful criticism of the Miltonic Adam is possible until the last trace of the simple, childlike Adam has been removed from our imaginations.* The task of a Christian poet presenting the unfallen first of men is not that of recovering the freshness and simplicity of mere nature, but of drawing someone who, in his solitude and nakedness, shall really be what Solomon and Charlemagne and Haroun-al-Raschid and Louis XIV lamely and unsuccessfully strove to imitate on thrones of ivory between lanes of drawn swords and under jeweled baldachins.[5]

Lewis Discusses the Fall in *The Problem of Pain*

In two chapters of his book *The Problem of Pain*, Chapter 4: "Human Wickedness" and Chapter 5: "The Fall of Man," Lewis opens with a statement similar to the one he made in his essay *Christian Apologetics*. As Professor Harold Bloom explains, until modern humans realize that they are sinful, "Christianity has little to say to them." For this reason, Lewis attempted to convince readers of their sinfulness. The second step was an attempt "to establish the doctrine of Original Sin without being literalistic about Adam and Eve", which could alienate many who would otherwise be receptive to Lewis' message.

Lewis' choice of words was not because he did not accept a literal Adam and Eve but, as Bloom explains, because he believes that in our modern Darwinian age much of his audience would not accept a literal

---

5. Italics added.

view of Adam and Eve.⁶ To deal with this real concern, Lewis phrases the Adam and Eve account as "a 'myth' in the Socratic sense, a not unlikely tale."⁷ As claimed by Joe Christopher, "What is significant in *The Problem of Pain* is that Lewis does not believe the Adam and Eve story can be taken seriously *by his audience* at a literal level in a Darwinian age."⁸ This is logical in view of the fact that Lewis once mentioned that he writes especially for atheists and agnostics.⁹ Bloom interprets Lewis' Adam and Eve account as a "myth":

> (or fabulous history) … which, at some point in time, god gave the gift of self-consciousness and awareness of the true, the good, and the beautiful…The rest of the account covers the Fall—that is, an act of self-will—and then dwindles into reflections on the event.¹⁰

As we have seen from Lewis' own writings, he made it very clear that he believed "man has fallen from the state of innocence in which he was created" and he therefore rejects "any theory which contradicts this" view.¹¹ As Professor Walsh wrote, to Lewis—as was also was true of orthodox Christianity—Christ was "the 'new Adam,' through whom humanity is to be restored to its original harmony with God. The final restoration will mean that man will once again have the same command over nature that Lewis attributes to the first men."¹²

Even if the Adam and Eve account was fiction, Walsh writes the words of Adam and Eve were ringing in the lead characters' ears of

---

6. Bloom, 2006, p. 118.
7. Bloom, 2006, p. 118.
8. Bloom, 2006, p. 118. Emphasis added.
9. Lewis, 1960, p. 49.
10. Bloom, 2006, p. 118.
11. Lewis, 2004b, p. 842. Letter dated September 23, 1944.
12. Walsh, 1949, pp. 105-106.

Lewis' books.[13] Walsh added that it was "not yet obvious that *all* theories of evolution contradict it."[14] A few years later it became very clear to Lewis that Darwinism *does* contradict it, and greatly so. As one of Lewis' characters, Aslan, states, "we came from Lord Adam and the Lady Eve" because all humans are "Sons of Adam and Daughters of Eve."[15]

Using modern terminology, *microevolution* does not contradict the Fall, but macroevolution or naturalistic evolution clearly does. Furthermore, Lewis argues that the unfallen Adam would not be inferior to us, but rather superior. The reason is because the powers of Jesus

> and the possible powers of unfallen Adam, are in rather a different category from either. Magic would be the artificial and local recovery of what Adam enjoyed normally: Which makes a difference. If Our Lord did His miracles *quâ* God and not *quâ* Man then the difference would be even greater.[16]

Again and again, Lewis wrote about how central the doctrine of Adam and Eve was to Christianity, which logically requires both "the doctrines of the Creation and the Fall" to be true. He added that some

> hazy adumbrations of a doctrine of the Fall can be found in Paganism; but it is quite astonishing how rarely outside Christianity we find…a real doctrine of Creation. In Polytheism the gods are usually the product of a universe already in existence… In Pantheism the universe is never something that God made. It is an emanation, something that oozes out of Him, or an appearance, something He looks like to us but really is not.… Polytheism is always, in the long run, nature-worship: Pantheism always,

---

13. Walsh, 1949, p. 44.
14. Lewis, 2004b, p. 842. Letter dated September 23, 1944.
15. Dickerson and O'Hara, 2009, p. 65.
16. Lewis, 2004b, p. 842.

in the long run, hostility to nature.[17]

In a detailed study of Lewis' beliefs, Brazier writes that the secular evolutionary "'religion' excluded miracles because it excluded the 'living God' of Christianity… Lewis proceeds to outline the roots of this popular religion in the Victorian Darwinism model of human evolution, which classified religion as a tribal human construct … this imagined history of religion is not true [because it comes from] the Darwinian evolutionary model."[18]

The Fall and Life in Outer Space

Lewis even speculated that if intelligent life was discovered in outer space, the possibility exists that "unfallen peoples may inhabit other planets."[19] As noted, the Fall doctrine was openly supported in several of Lewis' novels. For example, in his science-fiction novel *Perelandra*, Lewis' characters discussed the Fall.[20] A common notion in science-fiction novels was inhabitants of other planets would be cruel monsters, or at least monsters like some of H.G. Wells' creatures in his science-fiction novels.

These monsters often destroyed men who meant them no harm. In contrast, Lewis' secretary, Walter Hooper, noted that in Lewis' novels, fallen men went to planets where they came in contact with unfallen rational creatures. The Malacandra inhabitants in C. S. Lewis' novel *Out of the Silent Planet* did not even have words for sin or evil. They believed only that some persons were *bent*. A created being who was disobedient was referred to as "The Bent One."[21]

---

17. Lewis, 1970, p. 149.
18. Brazier, 2012, p. 184.
19. Deasy, 1958, p. 421.
20. Deasy, 1958, p. 422.
21. Lewis, 1967, p. 174.

Lewis also treated Adam as a real historical person in his private correspondence. He once wrote to his father that "since the Fall of Adam…[we] are but half men."[22] On March 17, 1953 he wrote to his friend, Italian priest St. Giovanni Calabria, explaining that a "necessary doctrine [held by both Catholics and Protestants is] that we are … closely joined together alike with the sinner Adam and with the Just One, Jesus."[23] And "the unity of the whole human race exists" because, Lewis wrote, we are all descendants of Adam.[24] Lewis also accepted creation *ex nihilo*, writing the "meaning of *creation*… I take… to mean … without pre-existing material (to cause both the form & matter of) *something pre-conceived in the Causer's thought.*"[25]

Hence, even in his early post-conversion life, Lewis did *not* "accept … the Darwinian concept of 'common descent with modification'" but, at the most, according to some interpretations, Lewis believed that this view could be an option for Christians. In other words, as noted above, the doctrine of the Fall, which was central to the thesis of his book *The Problem of Pain*, can be explained *even if* Darwinism was true. The contextual evidence, though, does not disagree with the conclusion that he accepted this view.

As Lewis explained, "I say 'evolution as popularly imagined.' I am not in the least concerned to refute Darwinism as a *theorem* in biology." He then acknowledged that there "may be flaws in that [Darwinian] theorem," and there

> may be signs that biologists are already contemplating a withdrawal from the whole Darwinian position, but I claim to be no judge of such signs. It can even be argued that what Darwin really accounted for was not the

---

22. Lewis, 2004b, p. 979.
23. Lewis, 2007, p. 306.
24. Lewis, 2007, p. 306.
25. Lewis, 2004b, p. 870.

origin, but the elimination of species ... For purposes of this article I am assuming that Darwinian biology is correct. What I want to point out is the illegitimate transition from the Darwinian theorem in biology to the modern myth of evolutionism or developmentalism or progress in general. The first thing to notice is that the myth arose earlier than the theorem, in advance of all evidence.[26]

As noted by Crowell, Lewis used the term developmentalism to refer to Darwinian evolution. Lewis also called for a change in the tendency to accept that "developmental change which we call Evolution, is justified by the fact that it is a general characteristic of biological entities" to change.[27]

---

26. Lewis, 1960, p. 101.
27. Williams, 2006, p. 88. See *That Hideous Strength*, 1974, p. 295.

The Piltdown Man. One of the most Famous hoaxes in history accepted because it fit the hopes and dreams of evolutionists. From author's collection.

24

Understanding Lewis' View of Evolution Continued

LEWIS RECOGNIZED that in his day "Evolution… is the *assumed* background of Christianity" because it was the dominant secular worldview, just as Ptolemaic Astronomy once was the *assumed* secular worldview of Christianity, but is no longer. Likewise, as Lewis studied the matter, he came to conclude that Darwinism was *not* in fact the background of Biblical Christianity.[1] Rather, Darwinism was (and is) the dominant *secular* worldview and the worldview held by academia, and those persons trained by the colleges including lawyers and, consequently judges, thus the courts and the legal system.

One critic of my position, Michael C., wrote: "C. S. Lewis clearly held that Adam and Eve were not spontaneously created, but came from a non-human anthropoid ancestor."[2] However, this is contradict-

---

1. Lewis, 2004b, p. 953.
2. See Product Review Page, https://amazon.com/product-reviews/1532607733/ref=cm_cr_arp_d_viewpnt_

ed by Lewis' statement that "The anthropoid has improved in the sense that he now is Ourselves."[3] By 'anthropoid' in this statement, Lewis appears to mean what he calls an "uninformed man" a "primitive," or person in terms of intellectual and or spiritual maturity, but a fully human man, not a pre-human ape-man, as Darwinism teaches.

A similar example is calling someone today a Neanderthal. In the popular idiom, the word "Neanderthal" is used as an insult to suggest that a person is deficient in intelligence, manners, or maturity. Lewis' use of the word anthropoid, as well as the term 'primitive man', could have been either "unfallen man or early fallen man", not an ape on its way to evolving into a man.[4] In Lewis' day, a major evidence purported to prove human evolution was the paleoanthropological fraud named "Piltdown Man", which was not exposed as fraudulent until 1953.[5] Java Man and Peking Man were also known but likewise controversial. That was the extent of the 'ape-to-man' evolutionary evidence.

What Lewis wrote elsewhere also argues for the view that he meant not an ape-man human evolutionary ancestor, but rather a culturally primitive, but fully-human, man. An example of Lewis' use of terms such as 'Neanderthal' to refer to a culturally primitive people is found in his inaugural lecture as Professor of Medieval and Renaissance Literature at Cambridge University. Lewis noted he was aware that some of his hearers didn't want "to be lectured on Neanderthal man by a Ne-

---

rgt?ie=UTF8&filterByStar=critical&reviewerType=all_reviews&pageNumber=1#reviews-filter-bar.

3. Lewis, 1960, p. 103.
4. Lewis, 2004b, p. 207.
5. De Groote, Isabelle, et al. 2016. New genetic and morphological evidence suggests a single hoaxer created 'Piltdown Man'. *Royal Society Open Science* **3**(8):160328, August. https://royalsocietypublishing.org/doi/pdf/10.1098/rsos.160328.

anderthaler, still less on dinosaurs by a dinosaur."[6] Lewis stated in his chapter on 'The Fall of Man' in *The Problem of Pain*, that the Fall "was transmitted by heredity to all generations, for it was the emergence of a new kind of man—a new species, never made by God, had sinned itself into existence."[7]

Lewis' Clear Statements Against Darwinism
Lewis also effectively argued against the idea of a Darwinistic brute, primitive ape-men idea taught by evolution.[8] After concluding that we live in what Lewis called an "absurd age", to illustrate the conclusion he used the example of a teacher whom he thought had been teaching evolution by explaining that, "'In the beginning was the Ape, from whom all other life developed – including such dainties as the Brontosaurus and the Iguanodon" when "Simple people like ourselves had an idea that Darwin said that life developed from simple organisms up to the higher plants and animals, finally to the monkey group, and from the monkey group to man." Lewis concluded that to accept this evolutionary view, "you need much more *faith* in science than in theology."[9]

Once fully convinced that the Darwinian evolution issue had a very important influence on theism, Lewis was motivated in some of his best writings to attack the idea with gusto. Even his fictional works reflected, not only Christianity, but also creationism. For example, in *The Chronicles of Narnia*, Lewis' character Aslan, who represents Christ, stated that the fixed categories of life existed in the "animal Eden; there is no evolution and certainly no blurring [from one to another]. It is all like Genesis."[10]

---

6. Deasy, 1958, p. 423.
7. Lewis, 1996d, p. 79.
8. Lewis, 1996d, pp. 73-74.
9. Lewis, 2004a, p. 227.
10. Bloom, 2006, p. 23.

An example of Lewis using the term evolution to illuminate some point which actually speaks *against* Darwinism is illustrated in the following quote:

> I should expect the next stage in Evolution not to be a stage in Evolution at all: I should expect the Evolution itself as a method of producing change will be superseded. And finally, I should not be surprised if, when the thing [evolution] happened, very few people noticed that it was happening. Now, if you care to talk in these terms, the Christian view is precisely that the Next Step has already appeared. And it is really new. It is not a change from brainy men to brainier men: it is a change that goes off in a totally different direction—a change from being creatures of God to being sons of God.[11]

C. S. Lewis used the term "Fall upward," for evolution in opposition to the Christian teaching of the Fall of Adam (the "Fall downward"). He invested much time and energy critiquing this view because he realized that this "Fall upward" idea was becoming an influential distortion of Christianity in his own day. Lewis thus rejected the Darwinian "Fall upward" of humans theory that Darwin argued for in his *Descent of Man*,[12] namely that humans began as immoral, selfish creatures, and that morality later evolved as we became social creatures. Darwin, Lewis believed, held this view in contrast to the view held by modern theologians who nevertheless regarded what they called the "Adam myth more as a story of Everyman than as a bit of ancient history."[13]

From what Lewis wrote, it is clear that he rejected this "Fall upward" view (which is a very good expression for evolution), and, at

---

11. Lewis, 1980, p. 186.
12. Darwin, Charles. 1871. *The Descent of Man and Selection in Relation to Sex*. London: John Murray.
13. White, 1969, p. 120.

the least, never openly endorsed it. The theologians, as is obvious from their words, often used "myth" to mean false, but Lewis used the word myth at times to mean "truth through symbolism."[14] West notes that, for Lewis, in contrast to Darwinism, evolution was a "purely biological theorem" that

> makes no cosmic statements, no metaphysical statements, no eschatological statements. Nor can Darwinism as a scientific theory explain many of the most important aspects of biology itself: "It does not in itself explain the origin of organic life, nor of the variations, nor does it discuss the origin and validity of reason." So what *can* the Darwinian mechanism explain according to Lewis?[15]

The answer is, not much. Lewis explained in his essay, "The Funeral of the Great Myth," given that because we were created and did not evolve, "we now have minds we can trust." Evolution attempts but fails to explain how "organic life came to exist," and "how a species that once had wings came to lose them [while evolution]…explains this by the negative effect of environment operating on small variations."[16] Thus, Lewis believed that

> Darwin's theory explains how a species can change over time by *losing* functional features it already has. Suffice to say, this is not the key thing the modern biological theory of evolution purports to explain. Noticeably absent from Lewis' description is any confidence that Darwin's unguided mechanism can account for the formation of fundamentally new forms and features in biology. Natural selection can knock out a wing, but can it build a wing in the first place? Lewis didn't seem to think so.[17]

---

14. White, 1969, p. 128.
15. West, 2012, p. 124.
16. Lewis, 1967, p. 86.
17. West, 2012, p. 124.

Consequently, natural selection can only explain the *loss* of structures, but not the gain of new structures. This negates the possibility of developmentalism, as in the evolution of fish to humans, for example. This fact creates major problems for methodological naturalism, says Lewis,

> because the *content* of evolution as a scientific hypothesis is much different from evolution as a popular myth. Inasmuch as evolution as myth superseded [went far beyond the facts of] evolution as a scientific hypothesis argues that the information trade-offs involved turned costly for scientific objects."[18]

In a letter dated August 1, 1949, to a Miss Breckenridge, Lewis wrote, "The doctrine of Evolution is that organisms have changed, sometimes for what we call (biologically) the better . . . *quite often for what we call (biologically) the worse.*"[19] In this statement, Lewis was quite correct.

We now know that around 99.9 percent of all mutations are near neutral (meaning slightly deleterious) or specifically deleterious. Thus, evolution is true, but going the wrong way; in Lewis' terms, "falling downward" and not "upward." This background helps us to understand Lewis' view of Darwinism was very different than that used by Darwinists today. It also helps us to understand Lewis' statement that "with Darwinianism [Darwinian evolutionism] as a *theorem* in Biology I do not think a Christian need have any quarrel."[20]

Lewis' Growing Problems with Evolution

Lewis' writings also document his growing problems with evolution.

---

18. Loomis and Rodriguez, 2009, p. 79.
19. Lewis, 2004b, p. 962.
20. Lewis, C. S. 1986. *Present Concerns.* New York: Harcourt Brace Jovanovich, p. 63. (Edited by Walter Hooper.)

On September 13, 1951 he wrote to creationist Bernard Acworth, who helped him understand the implications of evolution: "I must confess it [the information Acworth wrote on evolution] has shaken me: not in my belief in evolution, *which was of the vaguest and most intermittent kind*, but in my belief that the question was wholly unimportant."[21] West added that

> During the 1940s and 1950s, Lewis became more vocal about the looming dangers of what he called "scientocracy," the effort to hand over the reins of cultural and political power to an elite group of experts claiming to speak in the name of science. Lewis regarded this proposal as fundamentally subversive of a free society, and he worried about the creation of a new oligarchy that would "increasingly rely on the advice of scientists till in the end the politicians proper become merely the scientists' puppets."[22]

This concern became increasingly dominant in Lewis' later writings.

In the book *The Four Loves*, Lewis wrote that he disagreed with "those—and they are now the majority—who see human life merely as a development ... of animal life."[23] He added that evolution teaches an idea which he rejected as documented in the following quote: "all forms of behavior which cannot produce certificates [evidence] of an animal origin and [evidence] of survival value are suspect."[24] Lewis also objected to the behavioral implications of both evolution and natural selection, writing:

---

21. Lewis, 2007, p. 138; Ferngren and Numbers, 1996, p. 72.
22. West, 2012, p. 12.
23. Lewis, 1960, p. 90.
24. Lewis, 1960, p. 90.

Many schools of thought encourage us to shift the responsibility for our behavior from our own shoulders to some inherent necessity in the nature of human life, and thus, indirectly, to the Creator. Popular forms of this view are the evolutionary doctrine that what we call badness is an unavailable legacy from our animal ancestors."[25]

Professor West stressed that "Lewis took pains to emphasize that he was *not* 'anti-science'" but rather was

unequivocally opposed to *scientism*, the wrong-headed belief that modern science supplies the only reliable method of knowledge about the world, and its corollary that scientists have the right to dictate a society's morals, religious beliefs, and even government policies merely because of their scientific expertise.[26]

---

25. Lewis, 1996d, p. 60.
26. West, 2012, p. 12.

The progression from monkey to man is one of the most famous icons of evolution. It has been documented to be far more imagination than fact, as documented in the author's 390 page book. Author's book.

25

Lewis Becomes a Militant Anti-Darwinist

LEWIS WROTE MUCH about the "myth" of what we today refer to as Darwinism in his later writings, showing that his thinking developed well beyond his early speculation and possible equivocation about evolution. One example is when Lewis traced the history of science, he noted in his usual literary style that, "Darwin and Freud let the lion out of the cage," resulting in much harm to society. To illustrate this concern, Lewis wrote that the

> sciences long remained like a lion-cub whose gambols delighted its master in private; it had not yet tasted man's blood. All through the eighteenth century the tone of the common mind remained ethical, rhetorical, juristic, rather than scientific, so that Johnson could truly say, "the knowledge of external nature, and the sciences which that knowledge requires or includes, are not the great or the frequent business of the human mind."[1]

It is easy to see why Lewis concluded this, namely

---

1. Lewis, 1962, pp. 16-17.

> Science was not the business of Man because Man had not yet become the business of science. It dealt chiefly with the inanimate ... [until and] when Darwin starts monkeying with the ancestry of Man and Freud with his soul ... then indeed the lion will have got out of its cage.[2]

His concern was not the simple truism that the more fit were more likely to survive adverse circumstances, but his focus was "universal evolutionism," or what we refer to as the "common ancestry of all life from non-life."

Specifically, in Lewis' words, he was concerned about "The Great Myth" of evolutionism, namely the belief that "morality springs from savage taboos, adult sentiment from infantile sexual maladjustments, thought from instinct, mind from matter, organic from inorganic, cosmos from chaos."[3] Lewis opined that evolutionism is not only false, but is a belief proven to be very harmful to society. He added that this evolution of chaos to morality theory seems "immensely implausible, because it makes the general course of nature so very unlike those parts of nature we can observe."[4] Furthermore, Lewis argued that "The Great Myth" has influenced our common thinking. He asked on what grounds

> does "latest" in advertisements mean "best"? ... [it could be] that these semantic developments owe something to the nineteenth-century belief in spontaneous progress which itself owes something either to Darwin's theorem of biological evolution or to that myth of universal evolutionism.[5]

Lewis concluded by noting that the "obviousness or naturalness – which most people seem to find in the idea of emergent evolution

---

2. Lewis, 1962, pp. 16-17.
3. Lewis, 1996f, p. 137.
4. Lewis, 1996f, p. 137.
5. Lewis, 1962, p. 21.

– thus seems to be a pure hallucination."⁶ One of Lewis' main arguments against macroevolution as a whole, as summarized by Hooper, was as follows:

> We infer Evolution from fossils: we infer the existence of our own brains from what we find inside the skulls of other creatures like ourselves in the dissecting room. All possible knowledge, then, depends on the validity of reasoning. If the feeling of certainty which we express by words like *must be* and *therefore* and *since* is a real perception of how things outside our own minds really 'must' be, well and good. But if this certainty is merely a feeling *in* our own minds and not a genuine insight into realities beyond them—if it merely represents the way our minds happen to work—then we can have no knowledge.⁷

Furthermore, we are forced to infer evolution *indirectly* from fossils, not as direct evidence, just like we infer God from the creation and not as direct evidence. If evolution can be deduced from fossil evidence, then it is just as logical to infer God from the natural world.⁸ Actually, in reality, the fossils do not support evolution; consequently, we *can* infer God and His worldwide Flood judgment from the fossils themselves.

For these and other reasons, Lewis concluded that "a strict materialism refutes itself for the reason given long ago by Professor Haldane: 'If my mental processes are determined wholly by the motions of atoms in my brain, I have no reason to suppose that my beliefs are true … hence I have no reason for supposing my brain to be composed of atoms.'"⁹ Lewis concluded that science cannot determine truth by reason unless human reasoning is valid, and it

---

6. Lewis, 1962, p. 164.
7. Hooper, 1996, p. 601.
8. Beversluis, 2007, p. 191.
9. Hooper, 1996, pp. 601-602.

follows that no account of the universe can be true unless that account leaves it possible for our thinking to be a real insight. A theory which explained everything else in the whole universe but which made it impossible to believe that our thinking was valid, would be utterly out of court. For our theory would itself have been reached by thinking, and if thinking is not valid that theory would, of course, be itself demolished. It would have destroyed its own credentials. It would be an argument which proved that no argument was sound—a proof that there are no such things as proofs—which is nonsense.[10]

Lewis also opposed the ultimate basic foundation of Darwinism, namely chance, writing that if

the solar system was brought about by an accidental collision, then the appearance of organic life on this planet was also an accident, and the whole evolution of Man was an accident too. If so, then all our present thoughts are mere accidents—the accidental by-product of the movement of atoms. And this holds for the thoughts of the materialists and astronomers as well as for anyone else's.[11]

He added that if our

thoughts—i.e., of materialism and astronomy—are merely accidental by-products, why should we believe them to be true? I see no reason for believing that one accident should be able to give me a correct account of all the other accidents. It's like expecting that the accidental shape taken by the splash when you upset a milk jug should give you a correct account of how the jug was made and why it was upset.[12]

The problem of personifying evolution, he wrote, is that evolution is an abstract theory

---

10. Hooper, 1996, p. 602.

11. Lewis, 1984, p. 97.

12. Lewis, 1984, p. 97.

of all biological chances (as *sphericity* is of all spherical objects) … not an entity in addition to particular organisms… My point was that Butler, Bergson, Shaw, D. H. Lawrence etc. keep on talking as if it [evolution] *were* a thing…They call it *Life*. But *life* [highest common factor]…can't be alive any more than *speed* can move more quickly.[13]

Lewis also recognized that, in both the scientific and general literature, evolution is often spoken of as having human traits; including intelligence, wisdom, and foresight. In a letter Lewis wrote on May 4, 1961, he accused his former doctoral student, Dr. Alastair Fowler, of "biolatry," by which he meant personalizing evolution by making it into an intelligent force, adding "You talk of evolution as if it were a substance (like individual organisms) and even a rational substance or person. I had thought it was an abstract noun."[14] Lewis added that

> it is not impossible that, in addition to God and the individual organisms, there might be a sort of daemon, a created spirit, in the evolutionary process. But that view must surely be argued on its own merits? I mean we mustn't unconsciously and without evidence, slip into the habit of hypostatizing a noun.[15]

Lewis created in his novels both characters and situations that embody the concepts that he feared, and also those that he approved. A main character in three of his novels, Elwin Ransom, was a Christian who embraced the values of pity, kindness, honesty and respect for all individuals. In contrast, in the novels *Out of the Silent Planet* and *Perelandra*, the scientist Professor Weston believes that there exists no absolute truth, value or God. In this novel, "Weston is willing to sac-

---

13. Lewis, 2007, p. 1269.
14. Quoted in Poe, Harry Lee and Rebecca Whitten Poe, (editors). 2006. *C. S. Lewis Remembered: Collected Reflections of Students, Friends & Colleagues*. Grand Rapids, MI: Zondervan, p. 105.
15. Hooper, 1996, p. 496.

rifice anyone or anything to [achieve] his goal of perpetuating human life in the universe."[16] As Lewis in his own life did, Weston "became a convinced believer in emergent evolution... Science is no longer a substitute for religion; it has actually become a religion, or at least it imagines in its pride that it has become so.[17]

In the novel *That Hideous Strength*, the leaders of the scientific organization called the

> National Institute for Coordinated Experiments [N.I.C.E.] exhibit the ruthless disregard for people that Lewis feared would begin to appear in those who rejected traditional values; they would regard men as animals for experimentation, as some Nazis had so regarded the Jews. That his fears were not farfetched is seen by a brief glance at some developments in science since Lewis wrote these books.[18]

As C. S Lewis biographer, A. N. Wilson observed, "the general thrust of the book [*That Hideous Strength*] is that science is getting out of hand." An example Wilson notes is the statement that the "physical sciences, good and innocent in themselves, had, in Ransom's time, begun to be warped... maneuvered in a certain direction" to support conclusions that Lewis opposed, such as scientism and Darwinism.[19]

Darwinism Versus Variation Within the Genesis Kinds

The revolution in biology "from a devolutionary to an evolutionary scheme," Lewis explained, was "not brought about by the discovery of new facts," but rather by the "demand for a developing [evolving] world—a demand obviously in harmony both with the revolutionary and the romantic temper ... when it [belief in evolution] is full grown

---

16. Crowell, 1971, p. iv.
17. Pearce, 2013, p. 86.
18. Crowell, 1971, p. iv.
19. Wilson, 1990, p. 190.

the scientists go to work and discover the evidence."[20]

From both Lewis' nonfiction and fiction writings, Richard Cunningham has generalized that his main concern was that "Naturalistic empirical science must not be allowed to abrogate the right from limited empirical facts to project a whole philosophy about the nature of man and reality."[21] According to Cunningham, Lewis saw Logical Positivism as "the jumping-off place for naturalistic rationalism. The misuse of reason and the truncation of thought can be carried no further."[22]

In his "The Funeral of a Great Myth" essay, Lewis made one striking comment about Darwinian evolution. He concluded that "every age gets, within certain limits, the science it deserves," and our generation got the science it deserves, namely Darwinism and the baggage it brought with it, such as naturalism.[23] Lewis stressed that he did not "mean that these new phenomena [such as those observed in biology] are illusionary," but believed that Nature contains all sorts of phenomena that can be cherry-picked to "suit many different tastes" and theories, including evolution.[24]

Lewis here stressed that the doctrine of evolution "is certainly a hypothesis," although currently the best [naturalistic] hypothesis produced by biologists, it is a hypothesis nonetheless, not a fact. He added it may be shown "by later biologists, to be a less satisfactory hypothesis than was hoped fifty years ago."[25]

---

20. Lewis, C. S. 1964. *The Discarded Image: An Introduction to Medieval and Renaissance Literature.* Cambridge: Cambridge University Press, pp. 220-221.
21. Cunningham, 2008, p. 52.
22. Cunningham, 2008, p. 53.
23. Lewis, 1967, p. 85.
24. Lewis, 1967, p. 85.
25. Lewis, 1967, pp. 83, 85.

Lewis also argued for an eternal God by reasoning that "there was never a time when nothing existed; otherwise nothing would exist now," thus God, as Christianity teaches, always existed and will always exist.[26] Furthermore, "creation," as applied to human activity, is "an entirely misleading term" because we as humans can only *rearrange* what God has created because no *vestige* of real creativity *de novo* exists in us: "Try to imagine a new primary color, a third sex, a fourth dimension, or even a monster which does not consist of bits of existing animals stuck together ... we are [merely] recombining elements made by Him and already containing *His* meanings."[27]

In Clyde Kilby's biography of Lewis, after perceiving "the accusation that Lewis despises science" clarifies this charge by noting that what Lewis "actually dislikes is 'scientism' or the popular unthinking assumption that there is no truth other than truth revealed by the scientific method." To support this, Kilby quoted Lewis' essay "On Obstinacy in Belief,"[28] that

> distinguishes between scientific and Christian thought. Scientists, he says, are less concerned with "believing" things than simply finding this out. When you find something out, you do not any longer say you believe it, any more than you would say you believe the multiplication table.[29]

The fact is, materialistic scientists endeavor to escape from the problem of unbelief by turning

> belief into knowledge. Lewis defines belief as "assent to a proposition which we think so overwhelmingly probable that there is a psychological exclusion of doubt, though not a logical exclusion of dispute." The

---

26. Lewis, 2001a, p. 141.
27. Lewis, 2004b, p. 555. Letter dated February 20, 1943. Also Lewis, 1996, p. 203.
28. Kilby, 1964; the source is Lewis, 1960, *The World's Last Night*.
29. Kilby, 1964, pp. 175-176.

scientist himself holds "beliefs" of this sort concerning his wife ... which though not wholly subject to laboratory demonstration assume a large measure of evidence.[30]

He added that one might

rule out the view that any one passage taken in isolation can be assumed to be inerrant in exactly the same sense as any other.... because the story of the Resurrection is historically correct. That the over-all operation of Scripture is to convey God's Word to the reader (he also needs his inspiration) who reads it in the right spirit, I fully believe. That it also gives true answers to all the questions (often religiously irrelevant) which he might ask, I don't. The very kind of truth we are often demanding was, in my opinion, not even envisaged by the ancients.[31]

He added to this in a letter to Lee Turner:

"I myself think of it as analogous to the Incarnation – that, as in Christ a human soul-and-body are taken up and made the vehicle of Deity, so in Scripture, a mass of human legend, history, moral teaching etc. are taken up and made the vehicle of God's Word." [32]

This is recognized in the common understanding that no copy or translation, only the original autographs are inspired. Thus Bible manuscripts vary, but of the hundreds existing, the variations are very minor, mostly spelling changes. Furthermore, the Dead Sea Scrolls have verified that the versions existing today are extremely close to the

---

30. Kilby, 1964, pp. 175-176.
31. Letter to Clyde S. Kilby, dated July 5, 1959. In *The Collected Letters of C. S. Lewis, Volume 3: Narnia. Cambridge, and Joy, 1950-1963.*San Francisco, CA: HarperSanFrancisco, pp. 1044-1046.
32. Letter to Lee Turner, 19 July 1958 In *The Collected Letters of C. S. Lewis, Volume 3: Narnia. Cambridge, and Joy, 1950-1963.* San Francisco, CA: HarperSanFrancisco, pp. 960-961.

original, and no Christian teaching is affected. The most well-known examples are the two endings to Mark's Gospel; some translations include both endings, others include only one. None of these issues is a concern, even by those on the fringes of the creation issue, but, as Lewis notes, some problems still exist, a concern covered in some detail in commentaries, including those by creationists.[33]

The Bible is divided into history, proverbs, songs. parables and poetry, and it is critical to understand the difference. The Bible also includes many expressions no one takes literally, like the Earth's four corners mentioned over a dozen times, depending on the translation used. For example, Isaiah 11:12 states "He will raise a signal for the nations and will assemble the banished of Israel, and gather the dispersed of Judah from the four corners of the Earth."

For Young-Earth Creationists, the major concerns are: 'Is Genesis history?' and 'Is the length of the days of Genesis literal?' Thus, much detailed discussion has been expended on these questions, with good reason. Such discussion is important for determining the historicity of the book and if the creation days are literal (although many other numbers are not, a problem that exists in the translation of quantitative text from one language to another).

Furthermore, Lewis adds that it is wrong "to think that the obstinacy of Christians in their belief is like that of a poor scientist doggedly attempting to preserve a hypothesis [like Darwinism] although the evidence is against him."[34] He adds that their

> obstinacy is more like that of the confidence of a child who is told by its mother that to ease the pain from a thorn in its finger it must undergo

---

33. Morris, Henry.1976. *The Genesis Record: A Scientific and Devotional Commentary on the Book of Beginnings*. Grand Rapids, MI: Baker Books and Jonathan Sarfati, Jonathan. 2015. *The Genesis Account*. Creation Book Publications.

34. Kilby, 1964, pp. 175-176.

the additional pain of removal. That confidence rests not in a scientific demonstration concerning the mother but in confidence, even emotional confidence, in her as a person.[35]

The reality is, if an injured child acts on his or her "'unbelief,' and refuses to let its mother touch the finger, then no "mighty work can be done. Yet if it acts on its confidence the thorn will be got rid of and an increasing confidence in the mother will be established."[36] Likewise, "Christian doctrine requires us to put confidence in God who, being infinitely superior to us, will sometimes appear unreasonable to us, but in whom, as with the child and its mother, confidence yields the results promised."[37] He adds that, in Lewis' reply to Professor Haldane, Lewis argues

> the sciences are 'good and innocent in themselves,' though evil 'scientism' is creeping into them. And finally what we are obviously up against throughout the story is not scientists but officials. If anyone ought to feel himself libeled by this book it is not the scientist but the civil servant: and, next to the civil servant, certain philosophers.[38]

This is exactly what is occurring today.

---

35. Kilby, 1964, pp. 175-176.
36. Kilby, 1964, pp. 175-176.
37. Kilby, 1964, pp. 175-176.
38. Lewis, C. S. 1982, *C. S. Lewis on Stories and Other Essays on Literature*. New York: Harcourt Brace Jovanovich, p. 71.

Lewis Accepted the Modern Argument for Intelligent Design. This comes from the obvious claim that a watch design requires an intelligent designer. The watch example was made famous by William Paley. Picture credits Hannes Grobe/Hannes Grobe, CC BY 3.0 <https://creativecommons.org/licenses/by/3.0>, via Wikimedia Commons.

26

Lewis Increasingly Strident Opposition to Evolutionary Naturalism

LEWIS WAS MOST critical of the central thesis of Darwinism, its core foundational belief, namely naturalism. Darwinists attempt to show, through both science and reasoning based on naturalism, that our world and its inhabitants can be fully explained as the product of a mindless and purposeless system that evolved due to chance, time, and the operation of the laws of physics and chemistry. Lewis strongly opposed this worldview, using both reason and science. He taught that the something which exists beyond the material world is what we call the supernatural and that humans are on the borderline between

> the Natural and Supernatural. Material events cannot produce spiritual activity, but the latter can be responsible for many of our actions on Nature. Will and Reason cannot depend on anything but themselves, but Nature can depend on Will and Reason, or, in other words, God created Nature.[1]

1. Lewis, 1970, p. 276.

Furthermore, Lewis taught that what today is called Intelligent Design is manifest in the relation between nature and super-nature which becomes intelligible only

> if the Supernatural made the Natural. We even have an idea of this making, since we know the power of imagination, though we can create nothing new, but can only rearrange our material provided through sense data. It is not inconceivable that the universe was created by an Imagination [God] strong enough to impose phenomena on other minds.[2]

Although it appears that Lewis may have at times equivocated on evolution as early as 1925, his strong personal views against naturalism were clear from the date of his conversion until his death in 1963. A few weeks after the Scopes Trial concluded, Lewis wrote to his father about his change in academic fields.

Although he was glad for the academic change, he was grateful for what he had gleaned from his study of philosophy. A few years before he read Henri Bergson, Lewis wrote a letter to his father dated August 14, 1925, explaining his conclusion that Charles Darwin's theory of evolution and Herbert Spencer's theory of Social Darwinism were both built "on a foundation of sand." The context of Lewis' words was: "it will be a comfort to me all my life to know that the scientist and the materialist have not the last word: that Darwin and Spencer undermining ancestral beliefs [of theism] stand themselves on a foundation of sand; of gigantic assumptions and irreconcilable contractions an inch below the surface."[3] Of note is the fact that Lewis was still an atheist when he expressed these early doubts about Darwin and Darwinism.[4]

---

2. Lewis, 1970, p. 276.
3. In Lewis, 2004a. *The Collected Letters of C. S. Lewis. Volume 1: Family Letters, 1905-1931*. San Francisco, CA: HarperSanFrancisco, p. 649.
4. Lewis, 2004a, p. 213. Also letter of C. S. Lewis to his father, dated August 14, 1925 in *C. S. Lewis: Collected Letters* (2000). London: HarperCollins, Volume I, p. 649. (Edited by Walter Hooper.)

The date of his surrender to theism was four years later, in 1929.

Along this line, he also wrote that "One cause of misery and vice is always present with us in the greed and pride of men, but at certain periods in history this [problem] is greatly increased by the temporary prevalence of some false philosophy."[5]

This "false philosophy" was Darwinism that Lewis effectively questioned by opining how a brain that evolved could be trusted to determine if that which made it, evolution, were true. Lewis stressed that we must "infer Evolution from fossils" and that "all possible knowledge ... depends on the validity of reasoning," and thus, unless "human reasoning is valid no science can be true."[6]

He observed that, after studying the biological world, scientists had begun to study humans in the same way. Before this, scientists had *assumed* their own worldview was valid but now scientists'

> own reason has become the object [of evaluation]: it is as if we took out our eyes to look at them. Thus studied, his own reason appears to him as the epiphenomenon which accompanies chemical or electrical events in a cortex which is itself the by-product of a blind evolutionary process. His own logic, hitherto the king whom events in all possible worlds must obey, become merely subjective. There is no reason for supposing that it yields truth.[7]

Furthermore, because God created the natural world by creating

> it out of His love and artistry—it demands our reverence; because it is only a creature and not He, it is, from another point of view, of little account. And still more, because Nature, and especially human nature is fallen it must be corrected and the evil within it must be mortified. But

---

5. Lewis, 1967, p. 72.
6. Lewis, 1996c, p. 21.
7. Lewis, 1967, p. 72.

its essence is good.[8]

Reppert presented the logically sequential development of Lewis' thought in which he [Reppert] demonstrated—contrary to the dismissals of critics—that the basic thrust of Lewis' argument against Darwinism can effectively stand up to modern philosophical scrutiny.[9] For example, one of Lewis' more important works, *Miracles,* contains one of his most powerful critiques of the worldview that is at the foundation of both Darwinism and humanism, namely naturalism. Lewis observed that humanism was dominant as early as the 16th century and it "tended to be indifferent, if not hostile… to science.[10] Furthermore, he noted that even where Copernicus was accepted, "the change it produced was not of such emotional or imaginative importance as is sometimes supposed." Today, Copernicans' conclusions are so profound that his work is judged to have changed the world, producing the Copernican Revolution. Lewis observed that for "ages men had believed the earth to be a sphere. For ages… men realized that movements toward the center of the earth from whatever direction was downward movement."[11] This event, discussed in a chapter Lewis titled "New Learning and New Ignorance", is an example of rewriting history, called historical revisionism.

The purpose of historical revisionism was to support progressivism, a problem that Lewis was very concerned about. As Lewis correctly observed, the heliocentric idea did not produce the shock that "Darwinism gave to the Victorians or Freud to our own age." Specifically,

---

8. Lewis, 1970, p. 148.
9. Reppert, 2003.
10. Lewis, 1954, p. 2.
11. Lewis, C. S. 1954, "The Great Divide." *Christian History* 4(3):32. (An article written from his radio adaptation of his inaugural lecture as Professor of Medieval and Renaissance Literature given at Cambridge on 29 November 1954, pp. 2-3.)

in his *Miracles* book, Lewis wrote that adherents of naturalism assume that life "was not designed" because they do not accept that an Intelligent Designer exists. He then documented the fact that naturalism and naturalistic evolution are, at their core, atheism.[12]

Another concern Lewis often broached in his writing was the relationship between science and the government. One reason for this concern was possibly because Hitler and his government often abused science, specifically through social Darwinism and eugenics. As an example of his concern, Lewis wrote:

> 'Under modern conditions any effective invitation to Hell will certainly appear in the guise of scientific planning'—as Hitler's regime in fact did. Every tyrant must begin by claiming to have what his victims respect and to give what they want. The majority [of people] in most modern countries respect science and want [government] to be planned. And, therefore, almost by definition, if any man or group wishes to enslave us it will of course describe itself as 'scientific planned democracy.'[13]

Hitler believed that maintaining a "pure race" was necessary to insure the future of the Aryan race. An example Lewis provided to support his [Lewis'] view was the emergence of some political parties "in the modern sense—the Fascists, Nazis, or Communists." What distinguishes this modern political form "from the political parties of the nineteenth century is the belief of its members that they are not merely trying to carry out a program but are obeying an impersonal force: that [of] Nature, or Evolution, or the Dialectic, or the Race."[14]

Lewis admits that sometimes the actual state of affairs of people can "be so bad that a man is tempted to risk change even by revolutionary methods" because

---

12. Lewis, 2001a, pp. 27–28.
13. Lewis, 1982pp. 74-75.
14. Lewis, 1982, p. 78.

desperate diseases require desperate remedies and that necessity knows no law. But to yield to this temptation is, I think, fatal. It is under that pretext that every abomination enters. Hitler, the Machiavellian Prince, the Inquisition, the Witch Doctor, all claimed to be necessary.[15]

Lewis' novel *That Hideous Strength* was published at the end of WWII in 1945 when the horror of the concentration camps was made public. This novel mocks materialism: the main antagonists profess a belief in the idea that nothing exists apart from physical matter and energy.[16] Lewis, however, portrays the consequences of this worldview of rejecting God and His creation as a dystopian nightmare. In this novel, Lewis also made clear his contempt for Eugenics by using an analogy to express his concern: in a perfect society, natural trees will be replaced by lightweight aluminum trees so that no leaves will fall, and there will be "no twigs, no birds building nests, no muck and mess." After the environment is made perfect, next comes the people, including "sterilization of the unfit, liquidation of the backward races.... Selective breeding. Then real education... mainly psychological at first. But we'll get to biochemical conditioning in the end and direct manipulation of the brain."[17]

---

15. Lewis, 1982, p. 77.
16. https://en.wikipedia.org/wiki/Scientific_materialism" \o "Scientific materialism".
17. Lewis, 1996, p. 42.

Lewis referred favorably to the Jodrell Professor of Zoology and Comparative Anatomy, Professor D. M. S. Watson (1886-1973), at University College, London from 1921 to 1951.

27

The Funeral of the Great Myth

IN AN ESSAY TITLED "The Funeral of a Great Myth," Lewis explained why he regarded Darwinian evolution as the "great Myth of [the] nineteenth and early twentieth Century," a myth that he wanted to bury.[1] He explained why he called Darwinism, i.e., evolution, the great myth:

> because it is, as I have said, the imaginative and not the logical result of what is vaguely called 'modern science.' Strictly speaking, there is … no such thing as "modern science." There are only particular sciences, all in a stage of rapid change, and sometimes inconsistent with one another. What the Myth uses is a selection from the scientific theories—a selection made at first, and modified afterwards, in obedience to imagination and emotional needs. It is the work of the folk imagination, moved by its natural appetite for an impressive unity. It therefore treats its data with great freedom—selecting, slurring, expurgating, and adding at will.[2]

He even called "the Evolutionary Myth" a tragedy because of its

---

1. Lewis, 1967, p. 82
2. Lewis, 1967, pp. 82-83.

fruit, especially eugenics, and the destruction of the Christian church.[3] Lewis added that the "central idea of the myth is what its believers would call 'Evolution' or 'Development' or 'Emergence'" of higher life from lower life-forms.[4] In 1951, Lewis wrote that he was now inclined to think that evolution was "*the* central and radical lie in the whole web of falsehood that now governs" modern civilization.[5] Thus, he wrote "we must sharply distinguish between Evolution as a biological theorem and popular Evolutionism or Developmentalism which is certainly a Myth before proceeding to describe it and (which is my chief business) to pronounce its eulogy."

As noted, evolutionary naturalism, Lewis explained, is "not the logical result of what is vaguely called 'modern science,'" but rather is a picture of reality that has resulted, not from empirical evidence, but from *imagination*.[6] Furthermore, Lewis notes in *Mere Christianity* that most people believe in many things, including "evolution," on the basis of authority "because the scientists say so" and not on the basis of fact and scientific research and knowledge.[7] Lewis concluded that evolution theory emerged long before the necessary scientific research had been completed and, in making the myth, imagination has, and still does today, run ahead of the scientific evidence.[8]

Lewis stressed that the myth is bolstered by selecting facts from scientific theories, cherry-picking facts that are "modified ... in obedience to [the] imaginative and emotional needs" of the Darwinists.[9] Furthermore, Lewis argued, evolutionism infects minds as different as

---

3. Lewis, 1967, p. 84.
4. Lewis, 1967, p. 83.
5. Lewis, 2007, pp. 29-30, 138.
6. Lewis, 1967, p. 82.
7. Lewis, C. S. 1980, p. 63.
8. Lewis, 1967, p. 84.
9. Lewis, 1967, p. 83.

professors and media personalities such as Walt Disney, and "is implicit in nearly every modern article on politics, sociology, and ethics."[10] It has even infected English literature, such as that by H. G. Wells and Robert Browning.[11]

Lewis was especially concerned with the harm done by the "disingenuousness of orthodox biologists" due to their pushing evolutionary naturalism in the classroom and elsewhere.[12] Although Lewis admired much of the English literature that is based on the evolution myth, he had a major problem with its implications and its advocates forcing it on the public, writing that the evolution myth "has great allies. Its friends are propaganda, party cries, and bilge, and Man's incorrigible mind."[13]

His concern about natural selection is also shown throughout his novels, which include both good and bad spirits. Weston, a character in one of Lewis' novels, who is a non-theist scientist, "sees God and His angels as symbolic of the goal toward which evolution is striving and the Devil and his angels as the driving force behind the upward struggle [writing]; 'Your Devil and your God [Darwinism] are both pictures of the same force.'"[14]

Lewis Not Anti-Science but Anti-Scientism

Lewis' criticism was not of science in general "but of science without values, of science which makes itself God" such as Darwinism.[15] One of England's great modern scientists, evolutionist J. B. S. Haldane,

---

10. Lewis, 1967, p. 82.
11. Irvine, William. 1959. The Influence of Darwin on Literature. *Proceedings of the American Philosophical Society* **103**(5):616-628, October 15..
12. Ferngren and Numbers, 1996, p. 30.
13. Lewis, 1967, p. 112.
14. Lewis, C. S. 1996, *Perelandra*. New York: Scribner. First Edition 1944. p. 93.
15. Crowell, 1971, p. 27.

"charged that Lewis traduced scientists in *That Hideous Strength*."[16] Lewis replied to Haldane by charging that, if any of his novels "could be plausibly accused of being a libel on scientists it would be *Out of the Silent Planet*,"[17] which he admits is certainly an attack, but not on scientists, rather

> on something which might be called "scientism"—a certain outlook on the world which is usually connected with the popularization of the sciences, though it is much less common among real scientists than among their readers. It is, in a word, the belief that the supreme moral end is the perpetuation of our own species, and that this is to be pursued even if, in the process of being fitted for survival, our species has to be stripped of all those things for which we value it—of pity, of happiness, and of freedom.[18]

One of Lewis' novel characters zealously opposed Bergson's 'Creative Evolution' theory, which illustrates what

> Lewis had in mind when he wrote to Haldane of the danger of placing political power and scientific planning in the hands of those who felt that they had a mission to carry out. ...The horrors of the Inquisition, the Salem witch trials, the crushing of [Czech priest Jan] Hus's life and work in Bohemia and of [Reformer forerunner John] Wycliffe's in England, as well as communism's ruthless destruction of opponents in this century—all these are examples of this union of power and passionate belief which Lewis feared, but they lacked the power now known to science.[19]

One example where science has abused its power in the West is the movement to crush all efforts, both by Intelligent Design supporters, and all forms of theistic creationists, to do research opposing evolution

---

16. Crowell, 1971, p. 27.
17. Hooper, 1996, p. 76.
18. Hooper, 1996, pp. 76-77.
19. Crowell, 1971, p. 67.

and assume a place at the table of ideas. They are now largely ghettoized and cut off from mainline science.[20]

Lewis stressed that scientifically-minded Christians need to write, not just books on theology, but on the subject of true science:

> any Christian who is qualified to write a good popular book on any science may do much more by that than any directly apologetic work. The difficulty we are up against is this. We can make people (often) attend to the Christian point of view for half an hour or so; but the moment they have gone away from our lecture or laid down our article, they are plunged back into a world where the opposite position is taken for granted.[21]

Lewis then speculated that "this approach will not be very successful. What is successful is to attack the enemy's *line of communication*. What we want is not more books about Christianity, but more books by Christians on other subjects—with their Christianity *latent*."[22] Lewis italicized the word "latent" in his original book, knowing a direct (in-your-face or proselytizing) approach alienated people to the message. The subjects he listed needing to be addressed included elementary books on Geology, Botany and Astronomy; all three words Lewis capitalized.

His main concern was that the science fields "make the modern man a materialist; it is the materialists assumptions in all of the other books" that he was troubled about, the materialist worldview, namely Darwinism, which Ruse defines as a religion.[23] Lewis then made clear that the issue is "Primitive man" as taught in evolutionary science books as science, not history, i.e., evolution precursors leading up to

---

20. Bergman, 2012.
21. Lewis, 1970, p. 93.
22. Lewis, 1970, p. 93.
23. Ruse, 2017. p. 95.

the modern man, not the "Pre-historic" man which is part of documented history.[24] Lewis adds for many non-Christians "their real religion (i.e., faith in 'science')" blocks them from accepting Christianity.[25]

As documented, the common claim is that, because Lewis opposed scientism and Darwinism, he therefore was anti-science. For example, after noting the very real attractions of Lewis' trilogy, Deasy stated that one must consider an aspect of the trilogy that "for many readers proves an insuperable stumbling block. That is his total and unrelenting attack on science. It doesn't suffice to protest that that attack is against godless scientism."[26] Deasy added that, in Lewis' reply to Professor Haldane, Lewis argued:

> the sciences are 'good and innocent in themselves,' though evil 'scientism' is creeping into them. And finally what we are obviously up against throughout the story is not scientists but officials. If anyone ought to feel himself libeled by this book it is not the scientist but the civil servant: and, next to the civil servant, certain philosophers.[27]

This is exactly what is occurring today. In his trilogy books, Lewis wrote:

> there is no other science but the godless and dehumanizing, nor any other kind of scientist than those of the N.I.C.E.'s Power Elite, "dragging up from its shallow and unquiet grave the old dream of Man as God," men who, as is said of one of them, "had passed from Hegel into Hume, thence through Pragmatism, and then through Logical Positivism and out at last into the complete void."[28]

---

24. Lewis, 1970, p. 95.
25. Lewis, 1970, p. 95.
26. Deasy, 1958, p. 422.
27. Lewis, 1982, p. 71.
28. Deasy, 1958, p. 422.

Although Lewis was anti-Evolutionism or, using his terminology, an anti-developmentalist, he was not by any means anti-science, but believed that "all scientific theories are tentative and as dependent on changing presuppositions and climates of opinion as on new empirical data."[29]

Lewis wrote that evolution within "true science" is only a hypothesis about the changes that we observe every day in nature called microevolution, a view with which he had no problems. Conversely, in the popular mind, the myth is believed to be a "*fact* about *improvements*" in living organisms, a view that he strongly objected to.[30] The popular view of evolution involves common ancestry, life moving "upward and onward," an idea that Lewis also rejected. He added that, "if science offers any instances that seem to satisfy" the belief that life is evolving onward and upward, that idea "will be eagerly accepted. If it offers any instances that frustrate it [this view], they will simply be ignored."[31]

Lewis repeatedly stressed that the findings of science were not the final truth, but progress and develop, requiring one to be knowledgeable about some specific science topic.[32] Likewise, Christian apologists must answer the current negative attitude of the behavioral sciences against Christianity. This attitude is based on the fear Lewis observed that "the science which rejects or destroys traditional morality," was primarily in those disciplines which Lewis

> once called "pseudo-sciences," for example, sociology, and behaviorist and Freudian psychology (especially where scholars in these areas of study try to enter philosophy and religion without the necessary training and study). He makes a clear distinction between these groups and the

---

29. Quoted in Ferngren and Numbers, 1996, p. 31.
30. Lewis, 1967, p. 85.
31. Lewis, 1967, p. 86.
32. Dickerson and O'Hara, 2009. pp. 223-229.

pure sciences like chemistry and mathematics in the novel already mentioned.[33]

Lewis also noted that Christians must be cautious about using science to defend Christianity because theories which scientists adopted may be proven wrong years from now for the reason science:

> is in continual change and we must try to keep abreast of it. For the same reason, we must be very cautious of snatching at any scientific theory which, for the moment, seems to be in our favor. We may *mention* such thing; but we must mention them ... without claiming that they are more than 'interesting.'[34]

One must keep up-to-date because what is accepted today may be outdated tomorrow. For this reason, sentences beginning with the expression "science has now proved"

> should be avoided. If we try to base our apologetic on some recent development in science, we shall usually find that just as we have put the finishing touches to our argument science has changed its mind and quietly withdrawn the theory we have been using as our foundation stone. *Timeo Danaos et dona ferentes* [I fear the Danaans (Greeks), even when they bear gifts] is a sound principle.[35]

Lewis concluded that evolutionary naturalism is "immensely implausible, because it makes the general course of nature so very unlike those parts of nature we can observe."[36] Referring to the "chicken or the egg" question, Lewis wrote that moderns' acquiescence to "universal evolutionism is a kind of optical illusion, produced by attending exclusively" to the "does the chicken come from the egg or the egg from

---

33. Crowell, 1971, pp. 26-27.
34. Lewis, 1970, p. 92.
35. Lewis, 1970, p. 92.
36. Lewis, 1996f, p. 104.

the chicken" problem, adding, in support of Intelligent Design, that we "are taught from childhood to notice" how the oak tree

> grows from the acorn and to forget that the acorn itself was dropped by a perfect oak. We are reminded constantly that the adult human being was an embryo, never that the life of the embryo came from two adult human beings. We love to notice that the express engine of to-day is the descendant of the "Rocket"; we do not equally remember that the "Rocket" springs not from some even more rudimentary engine, but from something much more perfect and complicated than itself—namely, a man of genius. The obviousness or naturalness which most people seem to find in the idea of emergent evolution thus seems to be a pure hallucination.[37]

His point is that both of these examples are not evidence for evolutionary naturalism, but rather they are the result of innovation and creation due to Intelligent Design.[38] The growth of a tree from a seed was programmed by design, and, likewise, the invention and improvement of mechanical machinery was a result of design due to human intelligence. Neither evolved by random chance, damage to its components or mistakes, which comprise mutations, the mechanism of orthodox evolutionism. Rather, both came about by the opposite means, namely Intelligent Design. Lewis added that

> since the egg-bird-egg sequence leads us to no plausible beginning, is it not reasonable to look for the real origin somewhere outside [of the] sequence altogether? You have to go outside the sequence of engines, into the world of men, to find the real originator of the Rocket. Is it not equally reasonable to look outside Nature for the real Originator of the natural order?[39]

---

37. Lewis, 1996f, pp. 104-105.
38. Lewis, 1967, p. 90.
39. Lewis, 1970, p. 211.

This "Originator of the natural order" is the Intelligent Creator, called God in English. Lewis also wrote under the subtitle "all the guess work which masquerades as 'Science,'" that a scientific hypothesis "establishes itself [thus]... to use popular language, if you make the same guess often enough it ceases to be a guess and becomes a Scientific Fact."[40] He added that our world is being bombarded with the evolutionism Myth in hundreds of ways, including in government schools, colleges, universities, books, science films, museums, newscasts and the media as a whole.

Thus, Darwinism has become established in the West, not by evidence, but because it has been repeated *ad infinitum* in both the scientific and popular literature. Although Lewis once admitted that his knowledge of paleontology and geology was very limited, he was able to make insightful judgments on Darwinism based on viewing the entire Darwinian philosophy as a worldview, which it is.[41] One of Lewis' most telling illustrations of what he considered one of the most illogical aspects of evolution was presented under the heading "Evolution and Comparative Religion," where he discusses the basic problems of the "paleontological evidence" for evolution.[42]

Twisting the Facts

Lewis specifically objected to twisting facts so that they would fit into the evolution myth, such as changing microevolution from "a theory of change into a theory of improvement," thereby turning evolution into a worldview that explains all of creation, namely **macro**evolution, a philosophy which proposes that not

---

40. Lewis, C. S. 1944. *The Pilgrims Regress*. New York: Sheed and Ward, p. 36.
41. Lewis, 2004b, p. 848.
42. Lewis1944, p. 21.

merely terrestrial organisms but *everything* is moving "upwards and onwards." Reason has "evolved" out of instinct, virtue out of complexes, poetry out of erotic howls and grunts, civilization out of savagery, the organic out of inorganic, the solar system out of some sidereal soup.[43]

Lewis also discussed the "savage or Natural man... an ambivalent image ... but may be conceived as ideally innocent [as] Christians had depicted the naked Adam."[44] This comment helps us to understand what Lewis meant by the word "savage", which, from this context, does not imply a less evolved, primate ape man as implied by some, but rather an innocent and naïve pre-Fall man.

Lewis concluded that the theory of molecules-to-man evolution, that he called developmentalism, is not a deduction from the *data* using the accepted scientific method, but rather was a product of *imagination*. The human mind is all too easily convinced of its validity because many people *prefer* to believe that we, and our generation, are better than our parents and our parents better than theirs, and any theory which seems to reinforce this belief, such as evolution, appeals to us.[45]

In fact, humans tend to believe what they *want* to believe, or what their culture inculcates into them to believe, and the evidence is often secondary, if that. Lewis continues, noting the wonder of what evolution can do from nothing, that the status of zero, which can magically eventually turn into everything natural, is enormous. Furthermore, thanks to the wonders of evolution, he mocks the reasoning that "virtue, and civilization as we now know them are only the crude or embryonic beginnings of the far better things—perhaps Deity itself—in the remote future." Lewis explains:

in the Myth, "Evolution" (as the Myth understands it) is the formula for

---

43. Lewis, 1967, p. 86.
44. Lewis, 1954, p. 17.
45. Lewis, 1967, p. 82.

all existence. To exist means to be moving from the status of "almost zero" to the status of "almost infinity." To those brought up on the myth nothing seems more normal, more natural, more plausible, then that chaos should turn into order, death into life, and ignorance into knowledge. And with this we reach the full- blown Myth. It is one of the most moving and satisfying world dramas which have ever been imagined.[46]

Lewis asked how intelligent adults can believe that such foolishness as disorder can change into order. They do so because they were brought up to believe it, thus it seems as normal to them in the same way that those cultures which sacrifice maidens to the gods accept human sacrifice as normal. Then Lewis contrasts evolution with Christianity, writing:

> The doctrine of the Second Coming is deeply uncongenial to the whole of evolutionary or developmental character of modern thought. We have been taught to think of the world as something that grows slowly towards perfection, something that "progresses" or "evolves." Christian Apocalyptic offers us no such hope. The modern conception of Progress or Evolution (as popularly imagined) is simply a myth, supported by no evidence whatever."[47]

Note the clear statement of his conclusion here, writing "The modern conception of Progress of Evolution (as popularly imagined) is simply a myth, supported by no evidence whatever." By progress, he refers to molecules-to-man evolution, or from a common ancestor to modern man – *not*, for example, minor beak changes in finches.

As Corwin observed, "for C. S. Lewis, his problem with evolution was based on Biblical, philosophical, and logical grounds. He was aware that the great myth was antithetical to every Biblical doctrine.

---

46. Lewis, 1967, p. 86.
47. Lewis, C. S. 2012. *The World's Last Night and Other Essays*. New York: Mariner Books, pp. 100-101.

It was also antithetical to the 'Laws of logic' so crucial to empirical science."[48]

One of Lewis' main objections to "molecules-to-man evolution" was that "if the solar system was brought about by many accidental collisions [as evolutionists teach] then the appearance of organic life on this planet was also an accident, and the evolution of Man was an accident as well. If so, then all our present thoughts are [ultimately the result of] mere accidents—the accidental by-product of the movement of atoms."[49] In Lewis' words, if you believe that

> reality in the remotest space and the remotest time rigidly obeys the laws of logic, you can have no ground for believing in any [modern evolutionary] astronomy, any biology, any paleontology, and any archaeology. To reach the position held by the real scientists—which are then taken over by the Myth—you must—in fact, treat reason as absolute... how shall I trust my mind when it tells me about Evolution... The fact that some people of scientific education cannot by any effort be taught to see the difficulty [that evolution poses to reality] confirms one's suspicion that we here touch a radical disease in their whole style of thought.[50]

The Watson Quote

Lewis twice quoted Professor D. M. S. Watson (1886-1973), the Jodrell Professor of Zoology and Comparative Anatomy at University College, London from 1921 to 1951.[51] Watson was not a neophyte evolutionist, but had a wide interest and knowledge of fossils which he studied extensively in both the British Museum of Natural History in London, and on extended visits to South Africa, Australia, and the

---

48. Corwin, 2016, p. 52.
49. Lewis, C. S.. 1984. *The Business of Heaven*. San Diego, CA: Harcourt Brace Jovinovich,,p. 97. (Edited by Walter Hooper.)
50. Lewis, 1967, p. 89.
51. Lewis, 1967, p. 85.

United States.

Watson amassed a large collection of fossils, focusing on vertebrate paleontology, especially fossil reptiles. His acknowledged expertise earned him the position as curator of what is now the Grant Museum of Zoology. His early original work was on fossil plants and coal balls, a type of peat concretion varying in shape from an imperfect sphere to a flat-lying, irregular slab. Coal balls were formed when peat was prevented from being turned into coal by high amounts of calcite surrounding the peat. Given his background, what Watson wrote is very significant.

Specifically, quoting Professor Watson, Lewis wrote that Evolution "is accepted by zoologists not because it has been observed to occur or ... can be proved by logically coherent arguments to be true, but because the only alternative, special creation, is clearly incredible."[52] Lewis observed that this "would mean that the sole ground for believing it is not empirical but metaphysical—the dogma of an amateur metaphysician who finds 'special creation' incredible."[53]

Lewis then wondered: "Has it come to that? Does the whole vast structure of modern naturalism depend not on positive evidence but simply on an *a priori* metaphysical prejudice? Was it [evolution] devised not to get in facts but to keep out God?"[54]

Lewis answered this question in his writings, concluding that the statement by Professor Watson clearly was correct regarding the vari-

---

52. Quoted in Lewis, 1967, p. 85. Lewis quoted this from *The Nineteenth Century* (April 1943). The original is from D.M.S. Watson. "Adaptation." *Nature* 124(3119):231-234, 10 August 1929, p. 233. The article also appears in the *Report of the Ninety-Seventh Meeting British Association for the Advancement of Science* (Office of the British Association: London, 1929), pp. 88-99.

53. Lewis, 1967, p. 85.

54. Lewis, C. S. 1949. *The Weight of Glory*. New York: HarperCollins, p. 136. See also Lewis, 1996f, p. 104.

ation within the Genesis kinds. The Myth of Darwinism, Lewis then facetiously proclaims, regarding the myth of Darwinism, that "there is nothing hypothetical about it; it is basic fact; or, to speak more strictly, such distinctions do not exist on the mythical level at all." Rather, it is fact that is not to be questioned.[55]

Watson also incorrectly claimed that "whilst the fact of evolution is accepted by every biologist" a claim that ignores the many scientists who did not accept orthodox evolution then and also today.[56] After Watson speculated about why evolution was accepted, he admitted that "the mode in which it has occurred and the *mechanism by which it has been brought about are still disputable.*"[57] This section was not quoted by Lewis, but he would agree with it.

Lewis in his writings has attempted to understand the implications of Universal Evolution, writing that even

> if Evolution in the strict biological sense has some better ground than Professor Watson suggests—and I cannot help thinking it must—we should distinguish Evolution in this strict sense from what may be called the universal evolutionism of modern thought. By universal evolutionism I mean the belief that the very formula of universal process is from imperfect to perfect, from small beginnings to great endings, from the rudimentary to the elaborate, the belief which makes people find it natural to think that morality springs from savage taboos, adult sentiment from infantile sexual maladjustments, thought from instinct, mind from matter, organic from inorganic, cosmos from chaos.[58]

---

55. Lewis, 1967, p. 85.
56. Two of many classic examples include More, Louis Trenchard. 1925. *The Dogma of Evolution.* Princeton, NJ: Princeton University Press and Grasse, Pierre P. *The Evolution of Living Organisms.* New York, NY: Academic Press, 1977. For a list of close to 3,000 Darwin skeptics see. https://www.rae.org/essay-links/darwinskeptics/
57. Watson, D. M. S. 1929. "Adaptation." *Nature* 124(3119):231-234, August 10.
58. Lewis, 1949, pp. 136-137.

Lewis added that this "is perhaps the deepest habit of mind in the contemporary world." In other words, this is not only narrow-minded, but stubborn as well, because universal evolutionism, which he defines as "the belief that the very formula of universal process is from imperfect to perfect, from small beginnings to great endings, from the rudimentary to the elaborate, the belief which makes people find it natural to think that morality springs from savage taboos... mind from matter, organic from inorganic."[59]

This evolution is "immensely unplausible, because it makes the general course of nature so very unlike those parts of nature we can observe."[60] The reasons that the

> "modern acquiescence in universal evolutionism is a kind of optical illusion, produced by attending exclusively to the owl's emergence from the egg."[61]

Lewis concludes that "On these grounds and others like them one is driven to think that whatever else may be true, the popular scientific cosmology [developmentalism]... is certainly not." Then, expressing his concerns in laymen's language, Lewis explains that when he was taught in school to add a set of numbers, he had to 'prove my answer.' Likewise, the

> proof or verification of my Christian answer to the cosmic sum [evolutionary cosmology] is this. When I accept Theology I may find ... harmonizing with some particular truths which are imbedded in the mythical cosmology derived from science. But I can ... allow for science as a whole. ... [and because] Reason illuminates finite minds, I can understand how men should come, by observation and inference, to know a lot about the universe they live in.

---

59. Lewis, 1949, p. 137.
60. Lewis, 1949, p. 137.
61. Lewis, 1949, pp. 137-138.

Now comes the conclusion to the central matter, evolution versus creation: if "I swallow the scientific cosmology as a whole, then not only can I not fit in Christianity, but I cannot even fit in science."[62] One reason why is:

> If minds are wholly dependent on brains, and brains on biochemistry, and biochemistry (in the long run) on the meaningless flux of the atoms, I cannot understand how the thoughts of those minds should have any more significance than the sounds of the wind in the trees.[63]

Lewis adds that the final test is

> how I distinguish dreaming and waking. When I am wake I can ... account for and study my dream. The dragon that perused me last night can be fitted into my waking world. I know that there are such things as dreams; I know that I had eaten an indigestible dinner; I know that a man of my reading might be expected to dream of dragons. But while in the nightmare I could not have fitted in my waking experience. The waking world is judged more real because it can thus contain the dreaming world... For the same reason I am certain that the passing from the scientific points of view to the theological, I have passed from dream to waking. Christian theology can fit in science, art, morality...The scientific point of view cannot fit in any of these things, not even science itself. I believe in Christianity as I believe that the Sun has risen, not only because I see it, but because by it [the sun's light] I see everything else.[64]

Lewis even shared his experience of progressing from Darwinism to becoming a Christian, noting that the

> picture so often painted of Christians huddled together on an ever narrow strip of beach while the incoming tide of 'Science' mounts higher and higher corresponds to nothing in my own experience. That grand

---

62. Lewis, 1949, p. 139.
63. Lewis, 1949, pp. 139-140.
64. Lewis, 1949, pp. 139-140.

> myth which I asked you to admire ... is not for me a hostile novity [novelty] breaking in on my traditional beliefs. On the contrary, that cosmology [Darwinism] is what I started from. Deepening distrust and final abandonment of it [Darwinism] preceded my conversion to Christianity.[65]

This is one of several reasons why Lewis repeatedly returns to Darwinism and its problems in his writings, and not uncommonly covers some of the same points several times in his published writings. He continues, musing:

> Long before I believed Theology to be true I had already decided that the popular scientific picture [Darwinism]... was false. One absolute central inconsistency ruins it... The whole picture professes to depend on inferences from observed facts. Unless inference is valid, the whole picture disappears. Unless we can be sure that reality in the remotest nebula or the remotest part obeys the thought laws of the human scientist here and now in his laboratory... unless Reason is an absolute—all is in ruins.[66]

In discussions with his Oxford colleagues, Lewis was often frustrated because

> those who ask me to believe this world picture also ask me to believe that Reason is simply the unforeseen and unintended by-product of mindless matter at one stage of its endless and aimless becoming. Here is flat contradiction. They ask me at the same moment to accept a conclusion and to discredit the only testimony on which that conclusion can be based. The difficulty is to me a fatal one; and the fact that when you put it to many scientists, far from having an answer, they seem not to understand what the difficulty is, assures me that I have not found a mare's nest but detected a radical disease in the whole mode of thought from the very beginning. The man who has once understood the situation [must] re-

---

65. Lewis, 1949, pp. 134-135.
66. Lewis, 1949, p. 135.

gard the scientific cosmology [Darwinism] as being in principal a myth: though no doubt a great many true particulars have been worked into it.[67]

The problem is: how can evolution be universally accepted if the mechanism is still heavily disputed? Without a valid documented physical mechanism, the entire end goal of evolution – namely, as Darwin stated, "to murder God" by materialism – is negated. Darwin's breakthrough was a mechanism that he claimed could account for creation, namely natural selection.

Lewis went so far as to state indications existed that "biologists are already contemplating a withdrawal from the whole Darwinian position" on evolution.[68] One reason Lewis gave for this conclusion was the fact that "what Darwin ['s theory] really accounted for was *not the origin, but the elimination of species*" by natural selection.[69]

Darwinian Evolution Has Never Been Observed to Occur?

Some evolutionists have attempted to blunt the above quote by Watson (published in 1929 in *Nature*), which correctly stated that Darwinian evolution had never been observed to occur, but must be inferred from the fossil record, and its mechanism was still disputed; both still major problems. One response to Lewis using the quote by Professor Watson above was to call people like C. S. Lewis a crank. For example, Emeritus Professor of Geology, Steven Dutch, who wrote that, if you read critically enough, then in

> any site espousing some crank idea, whether creationism, geocentrism, or conspiracy theories, … you have a good chance of seeing some quote from a famous scientist. Lots of times the quotes are taken out of context, but in many other cases it's undeniable that the quote is just plain dumb.

---

67. Lewis, 1949, pp. 135-136.
68. Lewis, 1960, p. 101.
69. Lewis, 1960, p. 101. Italics added.

In some cases the scientists are blissfully unaware there are charlatans looking for scientific quotes to lend authority to their ideas; in other cases they seem to fancy themselves deep thinkers and unsung philosophers.

One of the best examples of this, Professor Dutch claims, is the Watson quote reproduced above:

> Probably no passages in all of science have been as widely quoted by creationists as these by British zoologist D. M. S. Watson ... Watson deserves a posthumous Ig Nobel Prize for doing more than any other mainstream scientist to advance creationism. Even allowing for the fact that this was written in 1929, it's scientifically illiterate.[70]

Of course, the Watson quote is not an example of scientific illiteracy, but an accurate observation of reality. Again, Lewis

> stressed Popular Evolutionism or Developmentalism [meaning Darwinism in Michael Ruse's terminology] differs *in content* from the evolution of the real biologists. To the biologist Evolution is a hypothesis. It covers more of the facts than any other hypothesis at present on the market and is therefore to be accepted unless, or until, some new proposal can be shown to cover still more facts with even fewer assumptions. At least, that is what I think most biologists would say.[71]

One of the best examples which illustrates C. S. Lewis' concerns about Developmentalism comes from the leader of Popular Evolutionism, Oxford professor emeritus Richard Dawkins, who wrote that the

> scientific principle that I wish everyone understood is Darwinian natural selection, and its enormous explanatory power, as the only known expla-

---

70. Steven Dutch, professor of Natural and Applied Sciences, University of Wisconsin-Green Bay. *Dumb Remarks by Scientists that Pseudoscientists Love* (2020). https://stevedutch.net/Pseudosc/StupidSci.htm

71. Lewis, 1967, p. 85.

nation of 'design'.⁷² The world is divided into things that look designed, like birds and airliners; and things that do not look designed, like rocks and mountains. Things that look designed are divided into those that really are designed, like submarines and tin openers; and those that are not really designed, like sharks and hedgehogs.⁷³

Dutch concludes that the means of determining design

> is that they are statistically improbable in the functional direction. They do something useful – for instance, they fly. Darwinian natural selection, although it involves no true design at all, can produce an uncanny simulacrum [an unsatisfactory imitation or substitute] of true design. An engineer would be hard put to decide whether a bird or a plane was the more aerodynamically elegant.⁷⁴

This is an example of speculation by evolutionists illustrating Lewis' concern that Darwinism is not a deduction from *data* using the accepted scientific method. Rather, it is a product of *imagination*. Professor Emeritus Dutch added the following claim which also supports Lewis' contrast of Developmentalism and the science of evolution based on experimentation:

> Unfortunately, Watson chose language that has been immensely useful to people who want to demolish science. Watson's central fallacy is one that has also been immensely useful to creationists: limiting the concept of "proof" in science to "experiment." Experimentation works fine in laboratory sciences but overgeneralizing it to all of science gives *carte*

---

72. About 80 percent of Americans, many of whom are well-educated (some are even scientists), disagree with this claim according to the 2019 Gallup poll on this subject.
73. Dutch, 2020. Note: Professor Dutch has labeled 80 percent of American's believers in pseudoscience (see footnote #755). https://stevedutch.net/Pseudosc/StupidSci.htm
74. Dutch, 2020.

blanche to people who want to attack historical or field sciences.[75]

Lewis certainly did not want to "demolish science" and neither do creationists. He was a strong supporter of science and spent his life teaching in two of the world's leading scientific institutions, Oxford and Cambridge University, working with many world-famous scientist colleagues. As Peterson wrote, although his (Lewis') academic field was medieval literature and the humanities, Lewis "wrote a surprising amount of material on science and its relation to both Christian faith and general culture."[76] Furthermore, Lewis "demonstrated remarkably good insight about the legitimate nature of science and critiqued distortions of science that serve ... secular philosophical" and sectarian goals, often called scientism, naturalism, or Darwinism.[77]

One of the best examples of Lewis' concerns about Developmentalism is by theoretical physicist Lawrence Krauss, who concluded that from his lifetime of studying the physical world, "the ultimate lesson from the story is that there is no obvious plan or purpose to the world we find ourselves living in. Our existence was not preordained, but appears to be a curious accident."[78] Krauss, a Jew by ethnicity, frequently quotes the Bible, which forces one to ask, "How can one possibly determine this conclusion from an empirical study of the planets, including the Earth, the stars, and the world of physics which shows fine tuning everywhere?" Of course, the answer is that one cannot.

---

75. Dutch, 2020.
76. Peterson, 2020, p. 125.
77. Peterson, 2020, . 125.
78. Krauss, Lawrence M. 2017. *The Greatest Story Ever Told – So Far: Why Are We Here?* New York: Atria Books, p. 4.

Trinil Man of Java, or Pithecanthropus, 500,000 years ago   Piltdown Man or Eoanthropus, 125,000 years ago   The "Cave Man" or Neanderthaler, 50,000 years ago   The Reindeer or Cró-Magnon man, 20,000 years ago

Restorations based on the skulls of four races of Old Stone Age men—the first three left no descendents

Extinct Races of Ape-Like Man
Tracing the Evolution of Man and the Apes from a Common Ancestor

All illustrations copyright, American Museum of Natural History

Lewis rejected the evolution of man from some ape-like ancestor as taught in the biology textbooks of the 1940s and 1950s. An example shown above from Scientific American was commonly used in biology textbooks. From author's collection.

28

The Canfield Letter

ONE EXAMPLE OF an attempt to prove that Lewis was an evolutionist who accepted the evolution of man from some ape-like ancestor, is the following:

> there is clear evidence that Lewis thought that universal common ancestry [of all life including humans] was perfectly consistent with the Christian faith. More than that, he was very clear that whether common ancestry is true with regard to human beings was a matter of indifference to him. In an unpublished letter to Joseph Canfield on Feb. 28. 1955, Lewis wrote: "I don't mind whether God made man out of earth or whether 'earth' merely means 'previous millennia of ancestral organisms.' If the fossils make it probable that man's physical ancestors 'evolved,' no matter."[1]

The claim that Lewis believed God using common descent to create the first human beings is irrelevant to Christianity is very misleading. In response to this claim, the letter itself will be quoted which shows that the letter actually supports the view presented in this book that

---

1. David Williams, from an internet postdated March 2020. See also Williams, David. 2012. "Surprised by Jack: C. S. Lewis on *Mere Christianity*, the Bible, and Evolutionary Science." *BioLogos*, December 10. https://biologos.org/people/david-williams

Lewis did *not* accept universal common descent of all life, especially of humans.[2] After thanking Mr. Canfield for his "kind and interesting letter," Lewis stated that what he is attacking in his story, *That Hideous Strength,* was

> not scientists but (what is growing in the real world) a kind of political conspiracy using science as its pretext. All tyrants avail themselves of whatever pretext is most popular in their age. As earlier tyrants made religion their pretext, of course modern ones make science theirs.

Lewis then adds that he is

> not a Fundamentalist in the strict sense—one who starts out by saying, in advance, "Everything we read here is literal fact." ... But *I often agree with the Fundamentalists about particular passages whose literal truth is rejected by many moderns.* I reject nothing on the ground of its being miraculous.

Of note is Lewis' statement that "I often agree with the Fundamentalists about particular passages whose literal truth is rejected by many moderns." He then correctly writes in reference to the Fall in Genesis that he doesn't

> see what the findings of the scientists can say either for or against it. You can't tell from looking at skulls & flint implements whether Man fell or not. But the question of the Fall seems to me quite independent of the question of evolution. I don't mind whether God made Man out of earth or whether "earth" merely means "previous Millennia of terrestrial organisms". If the fossils make it probable that man's physical ancestors "evolved," no matter. It leaves the essence of the Fall story intact.

Lewis did not say, "I am an evolutionist and believe in human evolution," but something very different, viz, "If the "fossils make it prob-

---

2. Unpublished letter from C. S. Lewis to Joseph Canfield, dated February 28, 1955. Original in the Wade Center Collection, Wheaton College (Wheaton, IL).

able that man's physical ancestors "evolved," ... It leaves the essence of the Fall story intact." His concern, which he makes clear, as documented in the previous chapters, is that

> One reason I don't attempt to fight on the purely biological issue is that I ... [would] have to study biology for 20 years before I could do it. Another is that one might waste time by attacking a position wh[ich] the scientists themselves might abandon tomorrow. The detection of the Piltdown forgery was fun, wasn't it?[3]

Lewis also correctly cautioned in the same letter the following:

> Sentences beginning 'Science has now proved' [which] should be avoided. If we try to base our apologetic on some recent development in science, we shall usually find that just as soon as we have put the finishing touches to our argument science has changed its mind and quietly withdrawn the theory we have been using as our foundation stone.[4]

Another way of expressing Lewis' conclusion, as penned by White, is that "the fundamentalist who believes in the devil 'hoofs, horns and all'... is closer to ultimate truth than the sophisticate who wholly denies the dimension of sin and evil in life."[5]

This is very different than the claims that David Williams and others make, and agrees with Lewis' writings in his book *God in the Dock* where he opined that if you practice some science it is very desirable that you keep up with the field, adding we need

> to answer the current scientific attitude towards Christianity, not the attitude which science adopted one hundred years ago. Science is in continual change and we must try to keep abreast of *it*. For the same reason, we must be very cautious of snatching at any scientific theory which, for

---

3. Lewis. Letter to Joseph Canfield dated February 28, 1955.
4. Lewis, 1970, p. 92.
5. White, 1969, p. 46.

the moment, seems to be in our favor. We may *mention* such things; but we must mention them lightly and without claiming that they are more than 'interesting.'⁶

Lewis added, seemingly with the Darwinism issue in mind, that sentences

beginning 'science has now proved' should be avoided. If we try to base our apologetic on some recent development in science, we shall usually find that just as we have put the finishing touches to our argument science has changed its mind and quietly withdrawn the theory we have been using as our foundation stone. *Inmeo Danaos et dona ferentes* is a sound principal.⁷

To understand Lewis' point here we must translate *Inmeo Danaos et dona ferentes*, in English *I fear the Greeks* (Danaans) *bearing gifts*, meaning don't trust an enemy even when he appears to bring you a gift. This refers to Virgil's (70 BC-19 BC) retelling of the Trojan Wars (13th-12th Century BC). When the Greek invaders left the shores of Troy, seemingly in defeat, they left behind a wooden horse as a gift for the people of Troy whom they had been fighting for many years. Despite the protestations of the priest, Laocoön, who advised the Trojans not to allow the horse into the city, the people brought it into the city, not knowing that hidden inside was a small band of Greeks who, once the city residents were asleep, left the horse and opened the gates of Troy to the invading Greek army who had quietly sailed back to Troy. Thus, Troy was finally defeated. Likewise, some science proof used to defend Christianity may hurt apologetics when it is found out later to be erroneous.

This paragraph reminds one of Lewis' statement "one might waste

---

6. Lewis, 1970, p. 92. Italics in the original.
7. Lewis, 1970, p. 92. Italics in the original. *Inmeo Danaos et dona ferentes* is translated, "I fear the Greeks (Danaans) even when they bear gifts."

time by attacking a position wh[ich]. the scientists themselves might abandon tomorrow. The detection of the Piltdown forgery [in 1953] was fun, wasn't it?"[8]

In conclusion, this is only one of many examples showing irresponsible distorting, or misquoting of Lewis to prove he was an evolutionist. The fact is, one major criticism of Lewis was that he held "unscientific, or fundamentalist attitudes and intentions."[9] While this may be true, in a study of Lewis, the Episcopalian priest Chad Walsh observed that "Lewis took more jabs at the liberal extreme in theology than at the fundamentalist extreme for the simple reason that he had less contact with fundamentalists in his Oxford environs" than other liberal theologians.[10]

---

8. Unpublished letter from C. S. Lewis to Joseph Canfield, dated February 28, 1955.Original in the Wade Center Collection, Wheaton College (Wheaton, IL).
9. White, 1969, p. 85.
10. White, 1969, p. 84.

Michael Ruse

Darwinism As Religion

What Literature Tells Us About Evolution

Lewis stressed evolution functioned as a worldview, actually a religion as documented by Professor Michael Ruse. Book cover.

29

Lewis Supports the Implications of a Creation Worldview

FOR MOST OF HISTORY it was widely believed that the universe had come into existence at some time in the past. This belief supported the Christian view that the universe was created by God. Modern physics supported Christianity when it proved that the "universe had a beginning," in contrast to the great materialistic systems of the past. In contrast, all of the major materialistic supporters believed the universe had always existed, as had self-existence of matter. Professor Whittaker said in the 1942 Ridell Lectures that

> 'It was never possible to oppose seriously the dogma of the Creation except by maintaining that the world has existed from all eternity in more or less its present state.' This fundamental ground for materialism has now been withdrawn. We should not lean too heavily on this, for scientific theories change. But at the moment it appears that the burden of proof rests, not on us, but on those who deny that nature has some cause

beyond herself.[1]

This cause beyond nature, Lewis concluded, is the Intelligent Designer we call God.

Lewis was also very aware of the fact, as documented by Michael Ruse in the first chapter of this book, that

> Darwinism has often functioned as a religion in itself. It certainly functioned that way for Thomas Huxley, "Darwin's bulldog." Evolution was a religion for politically varied figures such as playwright George Bernard Shaw, philosopher Fredrich Nietzsche, and composer Richard Strauss. Popular understanding of evolution actually owes much more to them than to Darwinian biologists.[2]

Furthermore, each of these writers had their

> own ideas about what evolution, specifically human evolution, was. However, Shaw got it right at least once. He explained in *Back to Methuselah* that "if this sort of [evolutionary] selection could turn an antelope into a giraffe, it could conceivably turn a pond full of amoebas into the French academy." That, precisely, is what Darwin and his successors believe, and what everyone who can be described as a non-Darwinist *disbelieves*.[3]

Nor was Lewis very impressed with the common argument that Darwinism is beyond doubt because it is the consensus of scientists, or what Lewis called the "climate of opinion." Lewis once even stated he finds it "almost impossible to believe that the scientists really mean what they seem to be saying," adding that one cannot "feel any certainty that some new scientific development may tomorrow abolish" some previously long held fact of science.[4]

---

1. Lewis, 1970, p. 39.
2. O'Leary, 2004, p. 64.
3. O'Leary, 2004, p. 64.
4. Beversluis, 2007, p. 146.

Whether the issue is evolution, or the current strident opposition to Intelligent Design, Lewis argued that it is not sufficient simply to acquiesce to the current "climate of opinion." In *The Problem of Pain*, he wrote that "I take a very low view of 'climates of opinion,'" noting that scientific discoveries are both often made by, and the "errors corrected by, those who *ignore* the 'climate of opinion.'"[5]

Lewis' "fears of what man might do to mankind" includes his study of certain trends in modern thought. The first trend of importance that he listed was the modernist view "that morality is relative and that moral standards have grown from mere impulses, from chemical reactions and responses which are in turn simply part of the irrational, blind development of organic life from the inorganic."[6]

The second trend Lewis feared

> is the idea that man will and should completely conquer nature, even human nature. The combined effect of these two ideas, Lewis feared, will spell the end of mankind as we know it … objective morality will have no guide but their own feelings and desires when they begin to control human nature.[7]

Specifically, Lewis had in mind the adverse moral results caused "by eugenics, by pre-natal conditioning, and by an education and propaganda based on a perfectly applied psychology."[8]

Atheist Recognizes Lewis' Rejection of Darwinism

Atheist Sunand Tryambak Joshi (he writes under the name S. T. Joshi) studied the classics at Brown and Princeton Universities. He has both B.A. and M.A degrees and is almost as prolific a writer as was C. S.

---

5. Lewis, 1962, p. 134.
6. Crowell, 1971, p. 12.
7. Crowell, 1971, p. 12.
8. Lewis, 1965, p. 72.

Lewis. Joshi recognized that Lewis rejected evolution and, in a fairly insightful discussion of Lewis' views on this topic, Joshi realized that Lewis divided all humans into two separate camps, the Naturalists (rationalists and evolutionists) and Supernaturalists (theists and creationists).

Professor Joshi then acknowledged that Lewis "pays lip service to evolution here and there... [but] it is clear that he was mightily uncomfortable with it, for a full acceptance of the theory would present insuperable difficulties to numerous facets of his theological reasoning."[9]

To show that Lewis clearly rejected evolution in spite of a few examples of "lip service" to the theory, Joshi relates an account that occurred in the late 1940s, mentioned above in more detail, when Lewis stated that "the first person he would wish to meet in Heaven was Adam." His host, Oxford Professor Helen Gardner, asserted that 'Adam' was "likely to be some barely articulate Neanderthaloid ape-man, to which Lewis snapped back: 'I see we have a Darwinian in our midst.'"[10]

Joshi claims that Lewis' rejection of evolution was due to the fact that Lewis did not have "the faintest comprehension of Darwin's theory of evolution."[11] If he did, Joshi believes, he would have rejected both his theology and his Christianity, and Lewis would have become an atheist and an evolutionist as Joshi did in college as a result of the influence of his college-professor parents.

Joshi also acknowledged that Lewis strongly rejected theistic evolution, a view he (Lewis) regarded as both poor theology and poor science. Joshi further outlines the convoluted "mental and emotional gymnastics" required to harmonize evolution with Lewis' creationism

---

9. Joshi, 2003, pp. 111-112.
10. Joshi, 2003, p. 112.
11. Joshi, 2003, p. 111.

theology.¹² An example Joshi cites to expose Lewis' "ignorance," which actually does show Joshi's ignorance, is as follows:

> It is painfully evident that Lewis was simply ignorant of the facts of science… his knowledge of the hard sciences appears to range from feeble to nonexistent. He is cataclysmically ignorant of the immense amount of research conducted even in his day by paleontologists, anthropologists, psychologists, biologists and historians that established unequivocally that primitive man had indeed slowly and gradually developed reasoning power over the course of the past two million years of evolution.¹³

On the contrary, the peer-reviewed evidence clearly shows that Joshi was wrong, as documented by Drs. Peter Line and Jerry Bergman.¹⁴

The Commercial World Welcomes the Myth

Lewis reasoned that advertisers welcome the Myth of Darwinism because it reinforced the belief that *the new model supersedes the old,* and therefore there must be an improved model that is more economical or better in other ways that replaced the old. Likewise, a powerful reason why politicians want to keep the myth alive is because they want us to believe that their new economic and fiscal packages are better than their previous ones. This was seen in both the 2016 and 2020 American election's stress on "change" advertised by both parties, especially the Democrats, as "change we can believe in."

Lewis stressed that, although the evolution Myth is "nonsense," "a man would be a dull dog if he could not feel the thrill and charm of it."¹⁵ Nonetheless, Lewis' advice is to "treat the Myth with respect"

---

12. Joshi, 2003, pp. 125-126.
13. Joshi, 2003, pp. 110-111.
14. Bergman, Jerry. et al. 2020. *Apes as Ancestors: Examining the Claims About Human Evolution.* Tulsa, OK: Bartlett Publishing.
15. Lewis, 1967, p. 93.

because the Myth

> gives us almost everything the imagination craves—irony, heroism, vastness, unity in multiplicity, and a tragic close. It appeals to every part of me except my reason. That is why those of us who feel that the Myth is already dead for us **must not make the mistake of trying to 'debunk' it in the wrong way**. We must not fancy that we are securing the modern world from something grim and dry, something that starves the soul. The contrary is the truth. It is our painful duty to wake the world from an enchantment.[16]

Lewis concludes that he grew up believing the evolutionary naturalism myth, that we humans are evolving and getting better with each passing year. He once felt—and continued to feel— the naturalism myth was "almost perfect grandeur":

> Let no one say we are in an unimaginative age: neither the Greeks nor the Norsemen ever invented a better story. Even to the present day, in certain moods, I could almost find it in my heart to wish that it was not mythical, but true. And yet, how could it be [true]?"[17]

After Lewis studied the Myth in detail, he concluded he could not accept it because he could not accept the claim that humans, and human reason, are "simply the unforeseen and unintended by-product of a mindless process [existing] at one stage of its endless and aimless becoming." The Myth itself negates the only grounds on which the Myth could possibly be true—that is reason.[18] He concluded that "For my own part, though, I believe it [the Myth] no longer."[19] Another example Lewis cites that negates the macroevolution Myth is the wonder of the human mind, noting that

---

16. Lewis, 1967, p. 93 Emphasis added.
17. Lewis, 1967, p. 88.
18. Lewis, 1967, 89.
19. Lewis, 1967, p. 93.

the Myth asks me to believe that reason is simply the unforeseen and unintended by-product of a mindless process at one stage of its endless and aimless becoming. The content of the Myth thus knocks from under me the only ground on which I could possibly believe the Myth to be true. If my own mind is a product of the irrational—if what seem my clearest reasoning are only the way in which a creature conditioned as I am is bound to feel—how shall I trust my mind when it tells me about Evolution?[20]

In Reppert's book noted above, *C. S. Lewis' Dangerous Idea* noted above, he [Reppert] documents the fact that Lewis effectively demonstrated that the Darwinian argument was circular. If materialism or naturalism were true, then scientific reasoning itself could not be trusted. Reppert adds, if we take our obligation to share our Christian faith seriously, Lewis showed that we will realize the Myth is a very real impediment for Christian evangelism today, both behind the scenes and out in the open.

Lewis notes that one evil of evolutionary naturalism is its abandonment of absolute moral, ethical, and logical standards. Thus, the conclusion from Naturalism that "moral codes are simply subjective values [that] evolved as people have developed through an evolutionary process from mindless matter is to say that we have no basis for choosing to call any idea good or bad or even to trust our own reasoning."[21] As Lewis illustrated in his novel, *That Hideous Strength*,

> many a mild-eyed scientist in pince-nez, many a popular dramatist, many an amateur philosopher in our midst, means in the long run just the same as the Nazi rulers of Germany. Traditional values are to be 'debunked' and mankind to be cut into some fresh shape at the will of some few lucky people in one lucky generation which has learned how to do it.[22]

---

20. Lewis, 1967, p. 89.
21. Crowell, 1971, p. 26.
22. Crowell, 1971, pp. 24-25. Quote from *Abolition*, p. 85.

Lewis described God as creating humans as described by Michelangelo.

30

Lewis Censors His Anti-Evolution Views

A MAJOR SOURCE of Lewis' anti-Darwinism views is in his manuscript titled "The Funeral of a Great Myth" which was not published until four years after his death in 1963. A second major source is his unpublished letters to Captain Acworth, president of the British *Evolution Protest Movement*.[1]

Lewis was far less open in public than in private about his opposition to evolution for several reasons. First, he was not a biologist and wanted to avoid openly confronting biologists because he felt, for good reasons, somewhat insecure in the natural science field. As a result, he attacked evolutionary naturalism in his writings largely from the perspective of logic and philosophy, fields in which he taught at Oxford and in which he was widely published.

Secondly, he realized that actively attacking evolution would produce much opposition to his person and writings and, as a result, would detract from his main work, Christian apologetics. He once noted that

---

1. Ferngren and Numbers, 1996. pp. 30-32.

"A great deal of my utility [in the apologist field] has depended on my having kept out of dog-fights between professing schools of 'Christian' thought."[2] He was also very aware of the often irrational attacks on those who rejected Darwinism as illustrated by the Scopes Trial and the well-documented, open and aggressive opposition to Christianity in Britain for much of the last century, especially in the university, as Lewis knew too well.[3]

Thirdly, very early in his career he may have flirted with theistic evolution, including common descent, and only after he explored the issue in some detail did he come to have major doubts about Darwinism. Some of his early comments have been uncritically interpreted as supporting evolution, but his published writings document that Lewis had major problems with evolutionism for most of his life. Consequently, Lewis "has been called an evolutionist by some and antiscientific by others."[4]

Finally, he did produce a well-reasoned book against naturalism titled *Miracles*, which was his major concern.[5] This book is still in print and is considered one of his most important works. The result of this conflict was that, for his entire life,

> Lewis remained reticent about speaking publicly on evolution. His three great apologetic works from the 1940s dealt with human origins only briefly and where absolutely necessary. Lewis believed he would do little good taking on a controversial subject in which he was not an expert. It is clear, however, from letters and essays that remained unpublished during his lifetime that Lewis did much reading and thinking in private about evolution. Until the 1950s, he tried to find a middle ground, accepting

---

2. White, 1969, p. 84.
3. Bergman, Jerry. 2018. *Censoring the Darwin Skeptics. How Belief in Evolution is Enforced by Eliminating Dissidents*. Southworth, WA: Leafcutter Press.
4. White, 1969, p. 84.
5. Lewis, 2001a.

the "biological theorem" while rejecting its "metaphysical statements." In the 1950s he grew more skeptical [of evolution]. Between 1944 and 1960, he corresponded privately with Bernard Acworth (1885-1963), one of Britain's leading anti-evolutionists.[6]

Nonetheless, Lewis did teach creationism in many of his books, even if only indirectly. For example, he wrote that "when God made space and worlds that move in space, and clothed our world with air, and gave us such eyes and such imaginations as those we have, He knew what the sky would mean to us. And since nothing in His work is accidental, if He knew, He intended."[7] Lewis was firmly convinced that "at least three things, joy, ethics, and human reason, ... could not have evolved [and in] ... the introductory chapter to *The Problem of Pain*, Lewis adds a fourth—religion."[8]

He also opined that "we finite beings may apprehend" God because "His glory [has been translated] into multiple forms—into stars, woods, waters, beasts, and the bodies of men," and it is in them—the evidence for the argument from design—that we see the proof for the existence of God.[9] The implications of this view are obvious, namely, if you cannot accept the view

> that the whole universe is a mere mechanical dance of atoms, it is nice to be able to think of this great mysterious Force rolling on through the centuries and carrying you on its crest. If, on the other hand, you want to do something rather shabby, the Life-Force, being only a blind force, with no morals and no mind, will never interfere with you like that trou-

---

6. Schultz and West, 1998, pp. 158-159.
7. Lewis, 2001a, p. 258.
8. Markos, 2003, p. 48.
9. Williams, Charles and C. S. Lewis. 1974. *Taliessin through Logres [and] The Region of the Summer Stars, by Charles Williams. And Arthurian Torso*. Grand Rapids, MI: Eerdmans, p. 291. (Introduction by Mary McDermott Shideler.)

blesome God we learned about when we were children.[10]

As noted, when he learned more about nature and the world, Lewis became increasingly hostile toward orthodox evolution. One example is a poem Lewis wrote about evolution titled *The Evolutionary Hymn* in which Lewis applied his consummate skills to mock evolution as a result of the fact that he not only "grew more skeptical" of evolution, but also more hostile to it. In the 1950s

> his doubts about evolution were being stimulated by the "fanatical twisted attitudes of its defenders." In 1957 he made his only public attack on the theory. In a poem titled "Evolutionary Hymn" he mocked evolution's pretensions to be a religion leading us "Up the future's endless stair."[11]

The complete poem is obviously a satire that blithely assumes a Darwinian view of the world and the inevitability of human progress. Old attic norms of good and evil are to be rejected since new ways and ideas are inherently superior (elsewhere Lewis called this attitude "chronological snobbery").[12] The poem, to be sung to the tune of the popular song *Joyful, Joyful We Adore Thee*, is as follows:

> Lead us, Evolution, lead us
>   Up the future's endless stair:
> Chop us, change us, prod us, weed us,
>   For stagnation is despair:
> Groping, guessing, yet progressing,
>   Lead us nobody knows where.
>
> Wrong or justice in the present,
>   Joy or sorrow, what are they
> While there's always jam to-morrow

---

10. Lewis, 1980, p. 35.
11. Schultz and West, 1998, pp. 158-159.
12. King, Don W. and Mike Perry in Schultz and West, 1998, pp. 158-159.

While we tread the onward way?
Never knowing where we're going,
  We can never go astray.

To whatever variation
  Our posterity may turn
Hairy, squashy, or crustacean,
  Bulbous-eyed or square of stern,
Tusked or toothless, mild or ruthless,
  Towards that unknown god we yearn.

Ask not if it's god or devil,
  Brethren, lest your words imply
Static norms of good and evil
  (As in Plato) throned on high;
Such scholastic, inelastic,
  Abstract yardsticks we deny.

Far too long have sages vainly
  Glossed great Nature's simple text;
He who runs can read it plainly,
  'Goodness = what comes next.'
By evolving, Life is solving
  All the questions we perplexed.

On then! Value means survival—
  Value. If our progeny
Spreads and spawns and licks each rival,
  That will prove its deity
(Far from pleasant, by our present
  Standards, though it well may be).[13]

It is significant that Lewis published this poem in the 1957 *Cam-*

---

13. Hooper, 1996, p. 176. This was originally written in 1954 in a letter to Dorothy Sayers.

bridge Review under the pseudonym "Nat Whilk."[14] Nat Whilk, which is actually the name of a number of Americans on Facebook, is Old English for "I know not whom" which is equivalent in English for "author unknown."[15] It was also only after Lewis' death that his two essays originally written in the 1940s, and containing his strong views against Darwinism and macroevolution, were published, namely "A Reply to Professor Haldane" and "The Funeral of a Great Myth."[16]

This fact also indicates that Lewis had strong objections to macroevolution decades before he died. The use of a pseudonym and delaying publications of his anti-evolution poem mocking Darwinism supports the view that he did not want to face the Darwin establishment's wrath which existed even in the 1940s, especially in the Ivy League Universities where he taught. This fact also is supported by the letters Lewis wrote to Acworth in 1951. In one letter, Lewis "politely declined to write a preface for one of Acworth's books, pointing out that, as a 'popular Apologist,'" he had to be careful because so many of his opposers were looking for "things that might discredit him."[17]

Some of Lewis' most cutting remarks about evolutionism were derived from his own academic background, i.e., literature, and he freely used literary phraseology to make his points. For example, an article he wrote in 1944 mocking evolution indicates that he had grave doubts about Darwinism long before the late 1950s. He wrote in his book *They Asked for a Paper* that we should consider the enormous claim of Christianity's

> chief contemporary rival— what we may loosely call the Scientific Out-

---

14. Schultz and West, 1998, p. 159.
15. Edwards, Bruce L. *C. S. Lewis: Life, Works and Legacy*. Volume 2: Fantasist, Mythmaker, and Poet. Westport, CN: Praeger Publishers, p. 36.
16. Reprinted in Chapter 7 of *Christian Reflections* by C. S. Lewis. Grand Rapids, MI: Eerdmans, pp. 82-93. (Edited by Walter Hooper.)
17. Schultz and West, 1998, p. 69.

look, the picture of Mr. Wells [a leading expositor of evolution in Lewis' day] and the rest [of the evolutionists]. Is it not one of the finest myths which human imagination has yet produced? The play [the evolution story] is preceded by the most austere of all preludes: the infinite void, and matter restlessly moving to bring forth it knows not what. Then, by the millionth millionth chance—what tragic irony—the conditions at one point of space and time bubble up into that tiny fermentation which is the beginning of life.[18]

He then added the observation that everything appears "to be against the infant hero of our drama," but in the end life somehow wins with

infinite suffering, against all but insuperable obstacles, it spreads, it breeds, it complicates itself: from the amoeba up to the plant, up to the reptile, up to the mammal. We glance briefly at the age of monsters. Dragons prowl the earth, devour one another and die.[19]

In the book *Miracles*, Lewis reasoned, as recounted by John G. West,

that the birth of modern science and its belief in the regularity of nature depended on the Judeo-Christian view of God as Creator: "Men became scientific because they expected Law in Nature, and they expected Law in Nature because they believed in a Legislator."[20] Nevertheless, Lewis thought that biology after Darwin provided potent fuel for turning science into a secular religion.[21]

According to what Lewis called this "fairy-tale," Lewis rejected the claim that humans evolved by the same process as did animals, thus

---

18. Lewis, 1962, pp. 154-155.
19. Lewis, 1962, pp. 154-155.
20. Lewis, 2001a, p. 106.
21. West, 2012, p. 21.

were animals. Lewis adds in his fairy-tale mocking Darwinism, that, as the weak spark of the first

> life began amidst the huge hostilities of the inanimate, so now again, amidst the beasts that are far larger and stronger than he, there comes forth a little naked, shivering, cowering creature, shuffling, not yet erect, promising nothing: the product of another millionth millionth chance. …He becomes the Cave Man with his club and his flints, muttering and growling over his enemies' bones, dragging his screaming mate by her hair (I could never quite make out why), tearing his children to pieces in fierce jealousy till one of them is old enough to tear him, cowering before the terrible gods whom he has created in his own image. But these are only growing pains. Wait till the next Act.[22]

Lewis continues with the next Act in what he called "this tale," in which the harm of the Myth becomes obvious in the form of eugenics, Freudian psychology, and communism. He wrote that Darwinism teaches that mankind is evolving into the true Man, learning to

> master nature. Science comes and dissipates the superstitions of his infancy [religion, creationism, and God]. More and more he becomes the controller of his own fate. Passing hastily over the present (for it is a mere nothing by the time-scale we are using), you follow him on into the future. See him in the last Act, though not the last scene, of this great mystery.[23]

The result was: "A race of demigods now rule the planet—and perhaps more than the planet—for eugenics have made certain that only demigods will be born, and psycho-analysis that none of them shall lose or smirch his divinity, and communism that all which divinity requires shall be ready to their hands."[24]

---

22. Lewis, 1962, p. 155.
23. West, 2012, p. 155.
24. Lewis, 1962, p. 155.

West indicated that Lewis believed modern science did not require the "kind of blind cosmic evolutionism promoted by H. G. Wells and company."[25] In order to detail what Lewis was opposing, he [Lewis] quoted Wells' evolutionary teaching, as printed in the 1928 edition of his *Short History of the World*, namely "All species of life upon Earth… are descended by slow continuous processes of change from some very simple ancestral form of life."[26]

Lewis was very concerned about the evil that would, and did, result by applying Darwinism to society. He wrote in 1943 that eugenics was part of the false hope which Satan wishes us to have under the illusion that we can turn our Earth into Heaven: "So inveterate is their appetite for Heaven that our best method, at this stage, of attaching them to earth is to make them believe that the earth can be turned into Heaven at some future date by politics or eugenics or 'science.'"[27]

Lewis ends the story with the depressing note regarding where the evolution Myth eventually leads – to orthodox atheism and the end of all life and the universe:

> If the myth stopped at that point, it might be a little bathetic [meaning used so often that the topic has lost interest]. It would lack the highest grandeur of which human imagination is capable. The last scene reverses all. We have the Twilight of the Gods. All this time, silently, unceasingly, out of all reach of human power, Nature, the old enemy, has been steadily gnawing away. The sun will cool—all suns will cool—the whole universe will run down. Life (every form of life) will be banished, without hope of return, from every inch of infinite space. All ends in nothingness, and *universal darkness covers all.*[28]

---

25. West, 2012, p. 21.
26. Lewis, 1967a p. 302. The book Lewis quoted is *A Short History of the World*. London : Labor Pub. Co., 1924
27. Lewis, C. S. 1943. *The Screwtape Letters*. New York: Macmillan, p. 144.
28. Lewis, 1962, pp. 153-156.

He summarized the Myth, noting that, although the pattern of myth appears to be one of the noblest that can be conceived, in fact it follows "the pattern of many Elizabethan tragedies, where the protagonist's career can be represented by a slowly ascending and then rapidly falling curve ... You see him [mankind] climbing up and up, then blazing in his bright meridian, then finally overwhelmed in ruin."[29] Lewis concluded that this world-drama appeals to every part of our psyche.

> The early struggles of the hero (a theme delightfully doubled, played first by [animal] life, and then by man) appeals to our generosity. His future exaltation gives scope to a reasonable optimism; for the tragic close is so very distant that you need not often think of it—we work with millions of years. And the tragic close itself just gives that irony, that grandeur, which calls forth our defiance, and without which all the rest might cloy. There is a beauty in this myth ... some great genius will yet crystallize it before the incessant stream of philosophic change carries it all.[30]

Lewis' opinion was this modern attitude stems mainly from the human attempt to study our own reason for belief, but this attempt is like taking "out our eyes to look at them. Thus studied, his own reason appears to him as the epiphenomenon which accompanies chemical or electrical events in a cortex which is itself the by-product of a blind evolutionary process. His own logic ... becomes merely subjective."[31]

Lewis adds that he believes "less than half of what" the evolution Myth "tells me about the past, and ... nothing of what it tells me about the future."[32] Aside from the evolution Myth, the only other plausible story of our origins is the Creation account of Christianity. This is the story Lewis spent much of his career writing about and extolling with

---

29. Lewis, 1962, pp. 155-156.
30. Lewis, 1962, pp. 153-156.
31. Lewis, 1967, p. 73.
32. Lewis, 1962, p. 156.

enormous success. He also recognized the fact that entropy is lethal to evolutionary theory:

> Evolution—even if it were what the mass of the people suppose it to be—is only (by astronomical and physical standards) an inconspicuous foreground detail in the picture. The huge background is filled by quite different principles: entropy, degradation, [and] disorganization.[33]

Furthermore, on the subject of evolution vs. entropy, Lewis wrote that the "march of all things is from higher to lower," not lower to higher, as both atheistic and theistic evolution teaches.[34] Furthermore, Lewis correctly stated, quoting J. B. S. Haldane who wrote, we are "inclined to regard progress as the rule in evolution. Actually it is the exception, and for every case of it there are ten of degeneration."[35] Lewis added that the Darwinism myth

> simply expurgates the ten cases of degeneration. In the popular mind the word 'Evolution' conjures up a picture of things moving 'onwards and upwards' and of nothing else whatsoever. And it might have been predicted that it would do so. Already, before science had spoken, the mythical imagination knew the kind of 'Evolution' it wanted. It wanted the Keatian and Wagnerian kind: the Gods ... If science offers any instances to satisfy that demand, they will be eagerly accepted. If it offers any instances that frustrates it, they will simply be ignored."[36]

Lewis correctly observed that scientists who attempt to publish articles "that frustrate it" (meaning to doubt or deny Darwinism) are destined to be either censored or ignored, thus being forced to start their own journals. Intelligent Design comes up often in his writing. Lewis, "on the subject of the spreading of man's corruption, wonders

---

33. Lewis, 1967, p. 58.
34. Lewis, 1970, p. 209.
35. Quoted in Lewis, 1967, p. 85.
36. Lewis, 1967, pp. 85-86.

if 'the vast astronomical distances may not be God's quarantine precautions.'"[37] Yet some pedants groundlessly conclude that C. S. Lewis would actually be opposed to the Intelligent Design ideology, writing "Lewis would have rejected intelligent design"![38]

I, however, am in agreement with "the C. S. Lewis Foundation, which is dedicated to advancing the renewal of Christian scholarship and artistic expression throughout the mainstream of our colleges and universities, and by extension, the culture at large."

---

37. Lewis, C. S. 1960. *Religion and Rocketry* essay, found in Lewis, 1960, *"The World's Last Night and Other Essays"*, p. 91.
38. Applegate, Kathryn. 2009. "C. S. Lewis on Intelligent Design." *In Pursuit of Truth | A Journal of Christian Scholarship*. http://www.cslewis.org/journal/cs-lewis-on-intelligent-design/view-all/.

Evolutionary biologists Professor J. B. S. Haldane was a critic of C S Lewis. Haldane was a professed socialist, Marxist, atheist, and secular humanist.

31

Lewis' Concern Fulfilled in the Movement Against Anti-Evolution

THE EVIL FORCES in Lewis' Space Trilogy novels are centered in a supposedly scientific organization—the *National Institute for Coordinated Experiments* (N.I.C.E.)—with a biologist and a psychologist as the important, albeit evil, characters. Some readers have regarded the book as an attack on science. Lewis replied to Professor J. B. S. Haldane's highly critical article of his novel that they were in no way an attack on science.[1]

Lewis answered Haldane's criticism by explaining specifically what he was attacking: "Firstly, a certain view about values"[2] "not scientific planning, as Professor Haldane had thought, but the kind of planned

---

1. Haldane, J.B.S. 1946. "Auld Hornie, F.R.S." *The Modern Quarterly*, pp. 32-40, Autumn.
2. Lewis, C.S. 1982. "A Reply to Professor Haldane," was published after C. S. Lewis' death and can be found, most recently, in the collection *On Stories: And Other Essays on Literature*, p. 78.

society which, first Hitler, and then European communists had instituted: 'the disciplined cruelty of some ideological oligarchy.'"[3] The Haldane article that critiqued Lewis' Space Trilogy was titled "Auld Hornie, F.R.S.", a name the Scots had given to the Devil. This article was published in the radical leftist magazine *The Modern Quarterly*.[4] Haldane claimed in this article that Lewis had many of his science facts wrong, suggesting that Lewis should have first brushed up on his science before he wrote the trilogy.

Haldane ignores the important fact that the three-part story is a freewheeling science-fiction novel based on the Genesis creation account, *not* empirical science. Haldane especially takes umbrage over the fact that there is only one respectable scientist in the three Space Trilogy books. All the rest are intolerant, or, at the least, science Fascists, a point that will be covered in much more detail in the following chapters. From these novels, Haldane concluded that Lewis had contempt for science, a claim Lewis responded to in detail, as also was covered above, especially as related to Darwinism. Haldane does admit that "the tale is told with very great skill, and the descriptions of celestial landscapes and of human and nonhuman behavior are often brilliant."[5]

Lewis correctly observed that every "tyrant must begin by claiming to have what his victims respect and to give what they want. The majority in most modern countries respect science and want to be planned ... therefore ... if any man or group wishes to enslave us it will of course describe itself as 'scientific planned democracy.'"[6]

Six years later, Lewis explained in a letter that "Where benevolent planning, armed with political or economic power, becomes wick-

---

3. Lewis, 1982, p. 80.
4. Haldane, 1946, pp. 32-40.
5. Haldane, 1946, p. 33.
6. Lewis, 1982, p. 80.

ed is when it tramples on people's rights for the sake of their good."[7] In *That Hideous Strength,* Lewis covers the problem of a few men controlling the rest of mankind, noting it appears that

> we now had the power to dig ourselves in as a species for a pretty staggering period, to take control of our destiny. If Science is really given a free hand it can now take over the human race and re-condition it: make man a really efficient animal ... Man has got to take charge of Man. That means, remember that some men have got to take charge of the rest.[8]

This attitude reveals "a lack of concern for individuals and their freedom, another concept that Lewis aggressively supported in his writings."[9] Lewis felt sure that naturalism and empiricism, by controlling men's views of the universe and each other, would eventually create a "union of applied science and social planning" that, with the power of government, would result in the loss of freedom and individuality. Consequently, he concluded that democracy was necessary to protect men from each other.

In response to Professor Haldane's claim that Lewis thought "the application of science to human affairs can lead to Hell,"[10] Lewis explained that it was not the application of science which he feared, but the application of science with the force of government.[11] This was true no matter what a person's choice might include. To make this idea clear to Haldane, who was then an active communist, Lewis further added

> I am a democrat because I believe that no man or group of men is good enough to be trusted with uncontrolled power over others. And the high-

---

7. Crowell, 1971, pp. 49-50.
8. Quoted in Dickerson and O'Hara, 2009, p. 223. Original in Lewis, 1996e. Originally published in 1945.
9. Crowell, 1971, p. 60.
10. Lewis, 1982, p. 80.
11. Crowell, 1971, p. 60.

er the pretensions of such power, the more dangerous I think it both to the rulers and to the subjects ... A metaphysic, held by the rulers with the force of a religion, is a bad sign. It forbids them, like the inquisitor [and later Nazis and Communists] to admit any grain of truth in their opponents, it abrogates the rules of ordinary morality, and it gives a seemingly high, super-personal sanction to all the very human passions by which, like other men, the rulers will frequently be educated.[12]

Note the 1949 statement by Professor Walsh about the creation-evolution controversy:

> One possible misconception can be quickly brushed aside. Lewis is not anti-scientific in a Fundamentalist sense. He is not troubled by the "conflict between science and religion" for the reason that his theology does not conflict with anything that science has so far discovered or is ever likely to discover. One cannot imagine him voting to prohibit the teaching of evolution in the schools of Britain.[13]

This review of Lewis' position on Darwinism has shown such a statement to be misleading, and actually irresponsible. In actuality, given his writings, one can more than imagine Lewis opposing the forcing of evolutionism (Darwinism/Neo-Darwinism/Punctuated Equilibria) on students today.

It is somewhat ironic that, in spite of the evidence that we read statements in *BioLogos* designed to foist evolution on the Christian community. This is exemplified by the following claim: "American Evangelicalism's infatuation with Lewis is in many respects somewhat odd. ... as we shall see, whose views on Scripture, Genesis, and evolution position him well outside of American Evangelicalism's standard theological paradigms."[14] To come to this conclusion, One member of

---

12. Lewis, 1982, p. 81.
13. Walsh, 1949, p. 129.
14. Williams, 2012 https://biologos.org/articles/surprised-by-jack-c-s-lewis-on-

BioLogos theologian David Williams relies on a few select quotes in some of Lewis' early writings, such as *The Problem of Pain* discussed in detail above. From these writings, he makes the amazing claim that throughout

> *The Problem of Pain Lewis* displays a remarkable degree of comfort with evolutionary theory, not least evolutionary accounts of human origins. A corollary of Lewis' acceptance of evolutionary theory, of course, is that death pre-existed humanity.

Williams quotes the following from *The Problem of Pain*:

> The origin of animal suffering could be traced, by earlier generations, to the Fall of man—the whole world was infected by the uncreated rebellion of Adam. This is now impossible, for we have good reason to believe that animals existed long before men. Carnivorousness, with all that it entails, is older than humanity.[15]

Williams responds to this statement by Lewis, writing that this quote shows

> how a broadly Darwinian picture of natural history may be compatible with a broadly Christian view of the world. For some, severing the link between the Fall of man and death's entry into the world, is anathema. But given Lewis' mere depravity view of the Fall, this evolutionary understanding of natural history creates no real problem for Christian faith.

However, this does not entail an evolutionary understanding, only that Lewis assumed that carnivorousness existed before the Fall. Another book would be required to consider the broad topic of the carnivorousness of, for example, insects, Platyhelminthes (flatworms), nematodes (roundworms), bacteria and even bacteriophages (phages or viruses that infect bacterial cells), and viruses in general, all which are

---

mere-christianity-the-bible-and-evolutionary-science.

15. Lewis, 1996d, p. 137.

parasites. These questions are still debated today even among Young-Earth Creationists.

Williams further quotes Lewis who wrote, "Perhaps a modern man can understand the Christian idea best if he takes it in connection with Evolution. Everyone knows about Evolution…: everyone has been told that man has evolved from lower types of life", ignoring the rest of Lewis' comments.[16] Williams attempts here to imply that Lewis endorses evolution by simply ignoring the rest of Lewis' comments on the subject.

---

16. Lewis, 1996d, p. 119.

Evolutionists John Dewey was one of the leading founders of the Humanists Movement. He was the first philosopher to recognize that Darwin's thesis not only required us to drastically change how we think about ourselves and the life around us, but it also required a markedly different approach to philosophy.
This picture was taken in 1902.

32

Letter on the Bible Book of Genesis from Janet Wise

Mrs. Janet Wise was a self-described "intelligent Fundamentalist," living in Mukalla Hadharaut, which was then part of the Eastern Aden Protectorate of Arab Sultanates, located in the southern part of the Arabian Peninsula. She was there possibly as a missionary. Mrs. Wise asked Lewis primarily about the problem of what she saw as the widespread disbelief in the Bible.[1] Lewis answered: "My own position is not Fundamentalist" if "Fundamentalism means accepting as a point of faith at the outset the proposition 'Every statement in the Bible is completely true in the literal, historical sense.'" The reason, Lewis explained, is that parables of Jesus and others are not "completely true in the literal, historical sense" but are stories designed to teach moral lessons. Lewis elaborated, explaining

> the same commonsense and general understanding of literary kinds wh

---

1. *The Canadian C. S. Lewis Journal.* Winter 1986. Number 53. Pp.1-3.

> [which] Wd [would] forbid anyone to take the parables as historical statements, carried a v. little further, wd [would] force us to distinguish between (1.) Books like *Acts* or the account of David's reign, wh [which] are everywhere dovetailed into a known history, geography, & genealogies (2.) Books like Esther, or Jonah or Job which deal with otherwise unknown characters living in unspecified periods, & pretty well *proclaim* themselves to be [parables similar to those of Jesus].

Lewis then added that other great Christians had some of the same concerns, such as

> Calvin [who] left the historicity of Job an open question and, from earlier, St. Jerome said that the whole Mosaic account of creation was done 'after the method of a popular poet'. Of course I believe the composition, presentation, & selection for inclusion in the Bible, of all the books to have been guided by the Holy Ghost. But I think He meant us to have sacred myth & sacred fiction [parables] as well as sacred history."[2]

Lewis' next words, ignored by evolution supporters, are: "Mind you, I never think a story unhistorical because it is miraculous. I accept miracles" which one would include the Genesis Creation account. Furthermore, this would have been an excellent place for Lewis to add that he viewed the Genesis Creation account as, not history, but as a parable intended only to teach us something about life, as is commonly claimed today by many theistic evolutionists. But he did not add this caveat.

This position agrees with Lewis' statement in his book *Reflections on the Psalms*, where Lewis wrote, "I have therefore no difficulty in accepting, say, the view of those scholars who tell us that the account of creation in Genesis is derived from earlier Semitic stories", a view

---

2. Letter to Janet Wise, dated October 5, 1955. *The Collected Letters of C. S. Lewis, Volume III: Narnia, Cambridge, and Joy, 1950-1963*, San Francisco: HarperSanFrancisco, pp. 652-653.

widely accepted by both Bible scholars and creationists.[3] For example, Young Earth Creationist Jonathan Sarfati writes that the "sources of Genesis are actually from eleven family documents headed by toledots." 'Toledot' is usually translated "these are the generations of..."[4]

Thus, Sarfati states, Moses was not the author but the *editor* of Genesis. The historic Church has accepted the inclusion of written accounts in Genesis as part of the canon. While Lewis' statement in the book *Reflections on the Psalms* seems unorthodox, Lewis, as an Oxford University scholar in medieval literature and mythology notes, in the field of literature, a myth is a traditional story which explains aspects of a people group's worldview or religious belief.[5] Lewis is not explicitly claiming, for example, that the Flood account in Genesis was the retelling of an earlier pagan story, the Epic of Gilgamesh, as is commonly claimed by those who dispute the Biblical record claim. A reading and a comparison of both the Biblical account and the Epic of Gilgamesh soon reveals the claim that the Biblical account was borrowed from the Epic of Gilgamesh to be blatantly false.

Lewis added that parables can be distinguished from historical sacred literature by several factors. The parables of Jesus, which forms about a third of his words in the New Testament, makes this clear with statements such as, "There once was a man who planted a vineyard" (from Mark 12; *The Passion Translation*).

Parables are an excellent means of illustrating profound truths, are easily remembered, the characters bold, and the symbolism rich. Lewis

---

3. Lewis, C. S. 1958. *Reflections on the Psalms*. New York: Harcourt Brace Jovanovich, p. 110.
4. Sarfati, Jonathan. 2015. *The Genesis Account. A Theological, Historical and Scientific Commentary on Genesis 1-11*. Powder Springs, GA: Creation Book Publishers, pp. 11-22.
5. Edwards, Bruce. 2007. *C. S. Lewis: Life, Works, and Legacy, Volume 1: An Examined Life*. Westport, CT: Praeger Publishers, p. 2.

closes his letter to Janet Wise with, "The basis of our Faith is not the Bible taken by itself but the agreed affirmation of all Christendom: to w[hic]h we owe the Bible itself." And for most of history, the Church understood Genesis as history written "after the method of a popular poet."

The St. Jerome story was checked in detail by Arend Smilde and published in a journal about C. S. Lewis titled *Journal of Inklings Studies*, published by the Oxford University C. S. Lewis Society. He concluded that "C. S. Lewis frequently quoted a testimony, supposed to be St. Jerome's, in which it is suggested that the biblical account of Creation was 'poetic' or 'mythical'. However, it seems Lewis had confused his authors and was ascribing to St. Jerome a passage actually written by the Renaissance humanist John Colet."[6] Furthermore, Lewis repeated the same claim at least six times, including in the book *Miracles* where he wrote,

> One seldom meets people who have grasped the existence of a supernatural God and yet deny that He is the Creator. All the evidence we have points in that direction, and difficulties spring up on every side if we try to believe otherwise. No philosophical theory which I have yet come across is a radical improvement on the words of Genesis, that 'In the beginning God made the heaven and Earth'. I say radical improvement because the story in Genesis – as St. Jerome said long ago – is told in the manner 'of a popular poet,' or as we should say, in the form of a folk tale. But if you compare it with the creation legends of other people—with all these delightful absurdities which giants to be cut up and floods to be dried up are made to exist *before* creation—the depth and originality of this Hebrew folk tale will soon be made apparent. The idea of creation in the rigorous sense of the word is here fully grasped.[7]

---

6. Smilde, Arend. 2014. C. S. Lewis, St. Jerome, and the Biblical Creation Story: The Background of a Recurring Misattribution. *Journal of Inklings Studies* 4(2):115-124, October, p. 115.

7. Lewis, 2001a, pp. 50-51.

Note the contrast between what Lewis judged as the superior Genesis account and the other inferior creation stories. Note, too, that folk tale does not mean false, nor does the word myth in this case. As reviewed above, both terms describe a class of accounts concerning origins called creation myths, and do not refer to the claim that they are false, but rather are part of a large anthological category.

St. Jerome, by using the phrase 'folk tale', was not implying the creation account is not true because St. Jerome, Lewis explains,

> was perhaps the greatest scholar among the Latin Church Fathers. Living and working in Bethlehem from 386 until his death, he made the Latin translation of the Bible known as the Vulgate, which was the standard Bible text for Western Christendom for the whole medieval period and in some ways authoritative as late as the 20th century. Surely if this mastermind of ancient Christianity held such a view of the biblical creation story, that is a thing worth recalling whenever it seems to be forgotten in modern discussions of the subject.[8]

Smilde elaborated, explaining

> anyone who has ever tried to trace Lewis' 'popular poet' quotation to a passage in Jerome's works must have concluded that it isn't there. ... Having checked a great many sources in Lewis' works over the years, I have found him not impeccable, but nevertheless fairly reliable as regards both letter and spirit of his countless quotations and allusions. Seeing how much he wrote and how much he quoted, and how uniquely successful he was in introducing modern readers to pre-modern authors, it is reasonable to grant him the right to a handful of blunders. It seems we must count this as one of them.[9]

The accuracy of the quote can easily be checked because all of Jerome's works are contained both in the *Online Library of Latin Texts*

---

8. Smilde, 2014, p. 115.
9. Smilde, 2014, p. 115.

and the *Patrologia Latina* database. The Latin phrase translatable as 'popular poet' is not found anywhere in either of these sources. The phrase does appear in the *Epistolae ad Radulphum*, written about 1,100 years after Jerome by John Cole.

The problem may have been solved in Lewis' book *English Literature in the Sixteenth Century*, described as '*The Completion of The Clark Lectures*, Trinity College, Cambridge 1944.' The book did not appear until 1954, but Lewis began writing it in the mid-1930s. This book contains the following passage:

> Colet ... has an important place in the history of Biblical studies and he is the most virulent of the humanists ... In the *Epistolae ad Radulphum* he himself allegorizes freely on the opening chapters of *Genesis* ... seeking a scientific or philosophical, rather than a moral or spiritual sense. It is one among many attempts made in this [i.e., 16th] century to reconcile the Mosaic account of the creation with the cosmological ideas of the day. In this difference between Colet's treatment of St. Paul and his treatment of *Genesis* there is inherent the recognition that the Bible contains books very different in kind. It was not exactly new – St. Jerome had allowed what we should now call the 'mythical' element in *Genesis* – but it was timely and useful.[10]

C. S. Lewis adds, "In his capacity as humanist we see Colet at his worst." The words of Colet are as follows:

> The firmament and heaven had been made first of all, in the day which he [Moses] calls first. But it was the design of Moses to touch on these more conspicuous objects afterwards in detail. And he does this after the manner of some popular poet, that he may the better study the spirit of simple-minded rustics; imagining a succession of events, and works, and times, such as could by no means find place with so great an Artificer.[11]

---

10. Lewis, 1954, pp. 159-160.
11. Colet, John. 1876. *Letters to Radulphus on the Mosaic Account of Creation, together with other treatises*, ed. J. H. Lupton (London: G. Bell) p. 9 (En-

In conclusion, perhaps Lewis carelessly checked the source, thus incorrectly attributed the phrase to Jerome instead of Colet. He remembered it, along with its mistaken attribution, for the rest of his life. As noted, Lewis repeated the claim at least six times in his writing. The latest was in a letter dated February 28, 1952 to a 'Mr. Canfield.' Lewis wrote, "The presence of an allegorical or mystical element in *Genesis* was recognized by St. Jerome. Origen held *Job* to be a moral fable not a history. There is nothing new about such interpretation." [12]

Note that the word 'mystical' is possibly a misrendering of the word 'mythical'.

Lewis continued, adding:

> I'm not a fundamentalist in the direct sense: one who starts out by saying, "Everything we read is literal fact." ... But I often agree with the Fundamentalists about particular passages whose literal truth is rejected by many moderns. I reject nothing on the grounds of its being miraculous. I accept the story of the Fall, and I don't see what the findings of the scientists can say either for or against it. You can't see for looking at skulls and flint implements whether Man fell or not. But the question of the Fall seems to me quite independent of the question of evolution. I don't mind whether God made Man out of the earth or whether "earth" merely means "previous material of some sort." If deposits make it probable that man's physical ancestors "evolved," no matter. It leaves the essence of the Fall itself intact. Don't let us confuse physical development with spiritual.[13]

This letter repeated much of what Lewis said before in other letters. It does add a few points, such as one related to a literal creation of Adam from the dust of the ground, namely: "I don't mind whether God made Man out of the earth or whether "earth" merely means "pre-

---

glish) and p. 170 (Latin).
12. Purtill, 1981, p. 83.
13. Letter dated February 28, 1952 to a 'Mr. Canfield.'

vious material of some sort." And "If deposits make it probable that man's physical ancestors 'evolved,' no matter". This is the most direct reference to human evolution, but again, note that he is not saying he accepts human evolution, only that *it may be proven in the future*. Notably, the ape to human fossil record has not only not been proven, but the evidence against ape to human evolution is overwhelming.[14]

Of course, the fossil evidence has not, at this date (2022) proven human evolution; far from it. Lewis' stand is quite clear from this interchange, but these facts do not seem to do much to quell the debate. For example, "Edgar Boss, a theological conservative, acclaimed Lewis as a champion of fundamentalists, even though he charged that Lewis was overly receptive to evolutionism and to liberal trends in biblical criticism."[15] If Boss were more aware of Lewis' views, he would not have expressed these concerns, at least his concerns about Lewis' receptiveness of evolution.

---

14. Bergman, Jerry. 2020 Apes as Ancestors: Examining the Claims About Human Evolution. Tulsa, OK: Bartlett Publishing. Co-Authored with Peter Line, PhD and Jeff Tomkins. PhD.
15. White, 1969, p. 83.

Lewis condemned Hitler and his racial theories.
In this photo he looks like a very angry man.

33

Some Conclusions

As HE READ AND THOUGHT on the matter, Lewis first left atheism and, increasingly became opposed to what Lewis called developmentalism and what is today called the "theory of evolutionism" or "Darwinism." In his earliest correspondence with Acworth,

> Lewis stated his willingness to accept any theory that does not contradict the fact that "Man has fallen from the state of innocence in which he was created." By 1951, however, he had begun to believe that Acworth might be right, that evolution was "the central and radical lie in the whole web of falsehood that now governs our lives."[1]

In brief, it appears that Lewis' conclusion

> about Darwinism as a biological theory changed over time. In the 1940s and 1950s, a friend tried to get him to join a protest movement against it. However, he refused, fearing that association with anti-Darwinists would damage his reputation as a Christian apologist. "When a man has become a popular Apologist," he explained, "he must watch his step. Everyone is on the lookout for things that might discredit him." However, privately, by 1959 he had become increasingly skeptical of Darwinism

---

1. Schultz and West, 1998, p. 69.

and concerned about its social effects, especially on account of "the fanatical and twisted attitudes of its defenders."[2]

When he was active as an apologist, Lewis was, at the very least, ambivalent about macroevolution, but always had problems with Darwinism even before he rejected atheism. As Ferngren and Numbers observed, "Lewis especially objected to the idea that human reason and an ordered universe could have arisen from the inorganic and irrational."[3] An example of his early ambivalence appears in a letter dated December 9, 1944, when he wrote to Bernard Acworth: "I am not either attacking or defending Evolution. I believe that Christianity can still be believed, even if Evolution is true. That is where you and I differ."[4] Insight as to why he said this follows in the same letter:

> Thinking as I do, I can't help regarding your advice (that I henceforth include arguments against Evolution in all my Christian apologetics) as a temptation to fight the battle ... on my *terrain* very unsuitable for the only weapon I have. Atheism is as old as Epicurus, and very few polytheists regard their gods as *creative*....[5]

His salutation revealed that he was on good terms with Acworth, and their correspondence lasted for almost two decades. This is the clearest claim Jack Lewis made in support of accepting evolution and, not unexpectedly, is often repeated by theistic evolutionists.[6] Lewis was cautious about openly attacking evolution in his early years because:

---

2. O'Leary, 2004, pp. 63-64.
3. Ferngren and Numbers, 1996, p. 31.
4. Quoted in Ferngren and Numbers, 1996, p. 30.
5. Lewis, 2004b, p. 633.
6. For example see the only negative review of my first book on C. S. Lewis (Bergman, Jerry. 2016. *C.S. Lewis: Anti-Darwinist*. Eugene, OR: Wipf & Stock). https://www.amazon.com/productreviews/1532607733/ref=acr_dp_hist_1?

Evolution was a creed so pervasive and so deeply held that even to appear to question it was to invite attack [from Darwinists]. For example, in a vitriolic article the Marxist geneticist J. B. S. Haldane accused Lewis of getting his science wrong and of traducing scientists in his works of science fiction.[7]

Lewis later came to realize that Darwinism *was a critical issue* "because evolution formed the basis of theories of philosophical naturalism like Haldane's, which had become the dominant secular worldview taught in our colleges and universities. Lewis agreed with Acworth regarding evolution 'as *the* central and radical lie in the whole web of falsehood that now governs our lives.'"[8]

This is one reason why, later in his life, "Lewis became increasingly critical of evolutionism and what he called 'the fanatical and twisted attitudes of its defenders.'"[9] Indeed, his works are among the most effective condemnation of both Scientism and Darwinism, and at the same time he presents one of the most effective philosophical defenses of Intelligent Design and creationism published in the last century. As Lewis explains, the popular thought

> that improvement is, somehow, a cosmic law: [is] a conception to which the sciences give no support at all. There is no general tendency even for organisms to improve. There is no evidence that the mental and moral capacities of the human race have been increased since man became man. And there is certainly no tendency for the universe as a whole to move in any direction which we should call 'good' [or more 'fit' as evolution teaches].[10]

Furthermore, from his experience, Lewis found that Christians

---

7. Ferngren and Numbers, 1996, p. 30.
8. Ferngren and Numbers, 1996, p. 32.
9. Ferngren and Numbers, 1996, p. 30.
10. Lewis, 1967, p. 58.

who accepted evolution, whom he called the "liberals," were often very *in*tolerant:

> In our days it is the "undogmatic" & "liberal" people who call themselves Christians that are most arrogant & intolerant. I expect justice & even courtesy from many Atheists and, much more, from ... Modernists [but] I have come to take bitterness and rancor as a matter of course [from Liberals].[11]

He made it very clear to laymen what he is referring to by the terms "evolution" and "liberals." Furthermore, it was

> obvious that what unites the Evangelical and the Anglo-Catholic against the 'Liberal' or 'Modernist' is something very clear and momentous, namely, the fact that both [Evangelicals and Anglo-Catholics] are thorough going Supernaturalists, who believe in the Creation, the Fall, the Incarnation, the Resurrection, the Second Coming, and the Four Last Things. This unites them not only with one another, but with the Christian religion as understood *ubique et ab omnibus* [everywhere and by all].[12]

This is exactly what creationists and Intelligent Design supporters stress today. In his later life, Lewis was one of the most effective anti-Darwinists of the last century.

Kuehn, in a review of Reppert, concluded that

> Reppert's work provides cogent support for the validity of Lewis' argument from reason, along with strong refutations of the many critiques leveled against it throughout recent decades. He shows that several of Lewis' critics do not take his evidences seriously on their own terms, instead preferring to dismiss them in favor of historical or biographical ("He wasn't even a professional philosopher!") considerations. He shows

---

11. Lewis, C. S. 1993. *Letters of C. S. Lewis. Revised and Enlarged Edition.* New York: Harcourt, p. 409.
12. Lewis, 1970, p. 336.

that such *ad hominem* Fallacies are particularly egregious in the analyses of Lewis' interactions with Elizabeth Anscombe.[13]

Lewis also feared a "Government in the name of science, opining that this is how tyrannies are produced, writing that in 'every age the men who want us under their thumb… will put forward the particular pretension which the hopes and fears of that age render most potent… It has been magic, it has been Christianity. Now it will certainly be science.'"[14] In support of Lewis' apprehension, Dickerson and O'Hara provided several examples of abuses by science. One was Francis Crick, one of the leading scientists today, who wrote in support of eugenics that certain people, presumably scientists, "should decide some people should have more children and some should have fewer… you have to decide who is to be born."[15]

Lewis also wrote that, although he firmly believed in God, he detested theocracy for the reason that "every Government consists of mere men and is, strictly viewed, a makeshift; if it adds to its commands 'Thus saith the Lord', it lies, and lies dangerously. On just the same ground I dread government in the name of science. That is how tyrannies come in."[16] Lewis explains his concern, namely that in

> every age the men who want us under their thumb, if they have any sense, will put forward the particular pretension which the hopes and fears of that age render most potent. They 'cash in'. …Perhaps the real scientists may not think much of the tyrants' 'science'; they didn't think much of

---

13. Kuehn, K. 2004. "Brains, Minds, and Unicorns: A Critical Review of Victor Reppert's C. S. Lewis' Dangerous Idea," p. 10. <http://www.ps.uci.edu/~kuehn/personal/reppert.htm>
14. Lewis, 1970, p. 315.
15. Dickerson and O'Hara, 2009, p. 226.
16. Lewis, C. S. 1958. "Willing Slaves of the Welfare State." *The Observer*, July 20. http://liberty-tree.ca/research/willing_slaves_of_the_welfare_state

Hitler's racial theories or Stalin's biology. But they can be muzzled.[17]

This conclusion has been shown true today by science's use of the courts and governments to effectively suppress criticism of Darwinism.[18] Professor Larson concluded that, "Lewis perceived science as the ultimate threat to freedom in modern society."[19] In this, Lewis proved correct, at least in reference to the hold that Darwinism and the science establishment have on society, our government and, especially, our courts.[20]

Lewis' concern was very clear: Darwinism leads to atheism. This is true in spite of the history of science that documents, "Men became scientific because they expected Law in Nature, and they expected Law in Nature because they believed in a Legislator", namely a creator God.[21] Lewis' concern was very clear: Darwinism leads to atheism.

Lewis added that "In most modern scientists this belief has died: It will be interesting to see how long their confidence in uniformity survives it. Two significant developments have already appeared—the hypothesis of a lawless sub-nature, and the surrender of the claim that science is true. We may be living nearer than we suppose to the end of the Scientific Age." Lewis' reference to "confidence in uniformity" is based on the fact that biological evolutionism is based on the doctrine of geological uniformitarianism. His reference to "lawless sub-nature" may have been related to Heisenberg's uncertainty principle which says

---

17. Lewis, 1958a.
18. Bergman,.2012; 2016. *Silencing the Darwin Skeptics*. Southworth, WA: Leafcutter Press; 2018.
19. Larson, Edward. 2012. "C. S. Lewis on Science as a Threat to Freedom." Chapter 3 in West, John G. (editor). 2012. *The Magician's Twin: C. S. Lewis on Science, Scientism, and Society*. Seattle, WA: Discovery Institute Press, p. 57.
20. Dickerson and O'Hara, 2009. pp. 226-227.
21. Lewis, 1947, p. 110.

that the "position and the velocity of an object cannot both be measured exactly, at the same time, even in theory. The very concepts of exact position and exact velocity together, in fact, have no meaning in nature."[22]

Of note is the fact that Heisenberg was working on developing an atomic bomb for Nazi Germany. This was the big weapon with which Hitler hoped to win WWII. One report concluded that "Heisenberg, Kurt Diebner, and Carl von Weiszacker were directly involved in the project" to build an atomic bomb.[23]

As noted in Chapter One, another writer who had a profound influence on Lewis was Sir Arthur Balfour, who summarized his anti-Darwinian conclusions into coherent arguments in his two books: *The Foundations of Belief* (1895) and *Theism and Humanism* (1915).[24]

Balfour argued that using Darwinism to support the idea that the human mind is a product of blind material causes was self-refuting: "all creeds which refuse to see an intelligent purpose behind the unthinking powers of material nature are intrinsically incoherent." [25] In the order of causation, they [these creeds] base reason on unreason and in the order of logic they employ conclusions that discredit their own premises.[26]

In his book *Theism and Humanism,* Balfour argued that, to maintain "our highest beliefs and emotions, we must find for them a congruous origin. Beauty must be more than accident. The source of morality must be moral. The source of knowledge must be rational."[27]

---

22. From https://www.britannica.com/science/uncertainty-principle
23. German Atomic Bomb Project. https://www.atomicheritage.org/history/german-atomic-bomb-project.
24. Balfour, 2000.
25. Raymond, E. T. 1920. *A Life of Arthur James Balfour*. Boston: Little Brown.
26. **Balfour, 2000, p. 147.**
27. **Balfour, 2000, p. 147.**

West concluded that Balfour produced an effective "critique of Darwinian and other materialistic accounts of human morality which 'he thought destroyed morality by depicting it as the product of processes that are essentially non-moral.'"[28] No wonder Balfour had such a major influence on Lewis.

Lewis was so very impressed and influenced by Balfour's argument, that he "named *Theism and Humanism* as one of the books that influenced his philosophy of life the most."[29] Lewis' debt to Balfour is also obvious in Lewis' book *Miracles: A Preliminary Study*, where he noted that modern materialists argue that the "mental behavior we now call rational thinking or inference must... have been 'evolved' by natural selection, by the gradual weeding out of types less fitted to survive."[30] Lewis then argued against this conclusion and "flatly denied that such a Darwinian process could have produced human rationality."[31] In support of this view, Lewis wrote that natural

> selection could operate only by eliminating responses that were biologically hurtful and multiplying those which tended to survival. But it is not conceivable that any improvement of responses could ever turn them into acts of insight, or even remotely tend to do so. Why not? Because "[t]he relation between response and stimulus is utterly different from that between knowledge and the truth known."[32]

Following Balfour, "Lewis held that attributing the development of human reason to a non-rational process like natural selection ends up undermining our confidence in reason itself."[33] This same logic was

---

28. West, 2012, p. 25.
29. West, 2012, p. 25.
30. Lewis, 2001a, pp. 27-28.
31. West, 2012, p. 25.
32. West, 2012, p. 28.
33. West, 2012, p. 25.

used by Notre Dame Professor Alvin Plantinga to defend theism. West adds that Lewis, in his book

> *Miracles* rejected a Darwinian explanation for the human mind because it undermined the validity of reason, he rejected a Darwinian account of morality because it would undermine the authority of morality by attributing it to an essentially amoral process of survival of the fittest.[34]

Lewis would today hardly be called a Darwinist, or even an evolutionist. The reason is because that which Orthodox Darwinism claimed to have explained purely due to the natural selection of mutations, including mind, body and morality, Lewis believed was due to divine creation by God. Furthermore, Richard Purtill, Professor of Philosophy at Western Washington University, compares the evolutionists' arguments with Lewis', writing that both

> Lewis and the "naturalists" can be seen as taking the trustworthiness of reason as a given and seeking an explanation for that agreed-upon fact. The naturalist's explanation is evolution; Lewis' explanation is that our reason is derived from the divine reason. What Lewis needs to argue, and indeed does argue indirectly, is that it is *overwhelmingly more probable that mind will be produced by a previously existing mind than by a process such as evolution*, which only selects characteristics favorable to survival under the conditions prevailing at a given time.[35]

The Anti-Darwinian Argument During Lewis' Lifetime

Some naïve readers might get the impression that Lewis knew as much about the origins debate as we know today. The scientific argument against evolution in Lewis' lifetime was very thin, and publications were very few compared to today. It is remarkable that Lewis managed to take his argument as far as he did. Henry Morris and John Whitcomb's *Genesis Flood* was published in 1961, only two years before

---
34. West, 2012, p. 25.
35. Purtill, 1981. p. 44. Emphasis added.

Lewis died, and we have no reason to believe that Lewis even heard either of the book or the modern *Creation Science Movement* (CSM). We can only wonder what his opinion of it would be if he did. It is likely that Lewis had available primarily Acworth's writings, and correspondence, plus Balfour's and Chesterton's criticisms as well as his own philosophical skepticism.

The *Evolution Protest Movement* (EPM) existed in his day, but was a small movement that had very limited influence. It began in 1932 and in 1980 became the Creation Science Movement. The Creation Movement in America was also small and limited to certain church denominations. The oldest existing creation organization in America, the *Creation Research Society* (CRS), was founded the same year Lewis died, in 1963 in Ann Arbor, Michigan. The Bible-Science Association was also started that year by Walter Lang. This ministry is now known as Creation Moments.

Nonetheless, even though he largely ignored the biological and paleontological arguments, Clive Staples Lewis effectively made one of the strongest, well thought-out cases against macro-evolutionism in his day and ours.

References

Aczel, Amir. 2007. *The Jesuit and the Skull: Teilhard de Chardin, Evolution, and the Search for Peking Man.* New York: Riverhead Hardcover.

Aeschliman, Michael D. 1998. *The Restitution of Man: C. S. Lewis and the Case Against Scientism.* Wm. B. Eerdmans Publishing Co

[See Evans, C. Stephen]

[See Fuller, Edmund]

Applegate, Kathryn. 2009. "C. S. Lewis on Intelligent Design." *In Pursuit of Truth | A Journal of Christian Scholarship.* http://www.cslewis.org/journal/cs-lewis-on-intelligent-design/view-all/.

Ayala, Francisco J. 2007. Darwin's greatest discovery: Design without designer. *Proceedings of the National Academy of Sciences* **104**:8567-8573, May 15.

Baen, James (editor). 1979. *Destinies.* Volume 1, no. 3, April-June. New York: Ace Books.

Balfour, Arthur James. 1915. *Theism and Human: Being the Gifford Lecture Series.* George H. Doran.

———. 2000. *Theism and Humanism: The Book That Influenced C. S. Lewis.* Edited by Michael W. Perry. Seattle, WA: Inkling Books.

Beavan, Colin. 2001. *Fingerprints: The Origins of Crime Detection and the Murder Case that Launched Forensic Science.* New York: Hyperion.

Bergman, Jerry. 2007. "Creative Evolution: An Anti-Darwin Theory Won a Nobel" *Impact* **409**:1-4, July.

——— and Joseph Calkins, M.D. 2009. "Why the Inverted Human

Retina is a Superior Design?" *CRSQ* **45**(3):213-224, Winter.

_____. 2011. "The Case for the Mature Creation Hypothesis." *CRSQ* **48**(2):168-177, Fall.

_____. 2012. *Slaughter of the Dissidents: The Shocking Truth About Killing the Careers of Darwin Doubters*. Revised version. Southworth, WA: Leafcutter Press.

_____. and Jeffrey P. Tomkins. 2012. "Is the human genome nearly identical to chimpanzee?—a reassessment of the literature." *Journal of Creation* **25**(4):54–60.

_____. 2016. *C.S. Lewis: Anti-Darwinist*. Eugene, OR: Wipf & Stock.

_____. 2018. *Censoring the Darwin Skeptics. How Belief in Evolution is Enforced by Eliminating Dissidents*. Southworth, WA: Leafcutter Press.

_____. 2019. *The "Poor Design" Argument Against Intelligent Design Falsified*. Tulsa, OK: Bartlett Publishing.

Bergson, Henri. 1944. *Creative Evolution*. New York: Random House. [See also: https://www.nobelprize.org/prizes/literature/1927/bergson/facts/]

Beversluis, John. 1985. "Beyond the Double Bolted Door." *Christian History* **4**(3):29, July.

_____. 2007. *C. S. Lewis and the Search for Rational Religion*. Amherst, NY: Prometheus Books.

Bishop, George. 1980. "Evolution: Blind Chance or God?" *The Teilhard Review* **15**(2):17-23, p. 23.

Bloom, Harold. 2006. *C. S. Lewis (Bloom's Modern Critical Views)*. New York: Chelsea House Publications.

Brazier, Paul. 2012. *C. S. Lewis: Revelation, Conversion, and Apologetics (C. S. Lewis: Revelation and the Christ, Book 2)*. Eugene, OR: Pickwick Publications (Imprint of Wipf and Stock Publishers).

Bredvold, L. 1968. The Achievement of C. S. Lewis, *The Intercollegiate Review* 4(2–3):1–7.

Burson, Scott and Jerry Walls. 1998. *C. S. Lewis and Francis Schaeffer: Lessons for a New Century from the Most Influential Apologists of Our Time.* Downers Grove, IL: InterVarsity Press.

The Canadian C. S. Lewis Journal. Winter 1986. Number 53. Pp1-3.

Chesterton, G. K. 1925. *The Everlasting Man.* London: Hodder & Stoughton.

Christensen, Michael J. 1979. *C. S. Lewis on Scripture.* Waco, TX: Word Books.

Colet, John. 1876. *Letters to Radulphus on the Mosaic Account of Creation, together with other treatises,* ed. J. H. Lupton (London: G. Bell), p. 9 (English) and p. 170 (Latin).

Collins, C. John. 2011. *Did Adam and Eve Really Exist? Who They Were and Why You Should Care.* Wheaton, IL: Crossway.

Corwin, Richard. 2016. *Creation Evolution and the Handicapped: Crushing the Death Image.* Bloomington, IN: WestBow Press division of Thomas Nelson.

Crowell, Faye Ann. 1971. *The Theme of the Harmful Effects of Science in the Works of C. S. Lewis.* M.A. Thesis, Texas A & M University.

Cunningham, Richard B. 2008. *C. S. Lewis: Defender of the Faith.* Eugene, OR: Wipf and Stock.

Darwin, Charles. 1871. *The Descent of Man and Selection in Relation to Sex.* London: John Murray.

_____. 1991. *The Correspondence of Charles Darwin: 1858-1859* – Vol 7. Cambridge University Press. Edited by Frederick Burkhardt.

Dawkins, Richard. 1986. *The Blind Watchmaker.* New York: W.W. Norton & Co.

_____. 2006. *The God Delusion.* Boston: Houghton Mifflin.

Deasy, Philip. 1958. "God, Space, and C. S. Lewis." *Commonweal*

68(16):421-425, July 25.

De Groote, Isabelle, et al. 2016. New genetic and morphological evidence suggests a single hoaxer created 'Piltdown Man'. *Royal Society Open Science* **3**(8):160328, August. https://royalsocietypublishing.org/doi/pdf/10.1098/rsos.160328.

Delfgaauw, Bermard. 1969. *Evolution: The Theory of Teilhard de Chardin*. New York: Harper & Row.

Dickerson, Matthew T. and David O'Hara. 2009. *Narnia and the Fields of Arbol: The Environmental Vision of C. S. Lewis*. Lexington, KY: University Press of Kentucky.

Dobzhansky, Theodosius. 1967. "Changing Man." *Science*, New Series **155**(3761):409-415, January 27.

Dorsett, Lyle W. 1985. "C. S. Lewis: A Profile of His Life." *Christian History*, **4**(3):6-11.

_____(editor). 1996. *The Essential C. S. Lewis*. New York: Touchstone. Includes C.S. Lewis' "De Descriptione Temporum", his inaugural address to Cambridge University.

Duriez, Colin. 2013. *The A-Z of C. S. Lewis: A Complete Guide to His Life, Thoughts and Writings*. Oxford, England: Lion Hudson.

_____. 2013a. *C. S. Lewis: A Biography of Friendship*. Oxford, England: Lion Books.

Dutch, Steven. 2020. *Dumb Remarks by Scientists that Pseudoscientists Love*. [Publisher?]

Edwards, Bruce L. 2007. *C. S. Lewis: Life, Works, and Legacy. Volume 1: An Examined Life*. Westport, CT: Praeger Publishers.

_____ 2007. *C. S. Lewis: Life, Works and Legacy. Volume 2: Fantasist, Mythmaker, and Poet*. Westport, CN: Praeger Publishers.

Evans, C. Stephen. 2004. "A Body Blow to Darwinist Materialism Courtesy of the Great C. S. Lewis." Review of *C. S. Lewis' Dangerous Idea*. The Book Service.

Ferngren, Gary B. and Ronald L. Numbers. 1996. "C. S. Lewis on Creation and Evolution: The Acworth Letters, 1944-1960." *The American Scientific Affiliation* 48:28-33, March.

Fowler, Alastair. 2003. C. S. Lewis: Supervisor. *Yale Review* **91**(4):64-80, October 1.

Frankl, Viktor E. 1969. "'Nothing But--' On Reductionism and Nihilism." *Encounter* **33**:54, November.

Fuller, Edmund. 1962. *Books with Men Behind Them*. New York: Random House.

Gardner, Martin. 1957. *Fads & Fallacies in the Name of Science*. New York: Dover.

Gibb, Jocelyn. 1965. *Light on C. S. Lewis*. London: Geoffrey Bles.

Glyer, Diana Pavlac. 2007. *The Company They Keep: C. S. Lewis and J. R. R. Tolkien as Writers in Community*. Kent, OH: The Kent State University Press.

Gould, Stephen Jay. 1980. *The Panda's Thumb*. Chapter 10: "Piltdown Revisited", pp.108-124. New York: W.W. Norton & Company.

Green, Roger Lancelyn and Walter Hooper. 1974. *C. S. Lewis: A Biography*. New York: Harcourt Brace & Company. Revised edition, 1994. New York: Harvest Books.

Gruber, Howard E. 1974. *Darwin on Man: A Psychological Study of Scientific Creativity*. Second Edition. Chicago, IL: The University of Chicago Press.

Haldane, J.B.S. 1946. "Auld Hornie, F.R.S." *The Modern Quarterly*, pp. 32-40, Autumn.

Hamilton, Virginia. 1988. *In the Beginning: Creation Stories from Around the World*. San Diego: Harcourt Brace Jovanovich.

Hart, Dabney. 1985. "Teacher, Historian, Critic, Apologist." *Christian History* **4**(3):21-24.

Hooper, Walter (editor). 1982. *Of Other Worlds*. New York: Harcourt

Brace Jovanovich. ("A Reply to Professor Haldane," pp. 69-79.)

_____(editor). 1996. *C. S. Lewis: A Companion & Guide*. London: HarperCollins Publishers.

_____(editor). 2004. *The Collected Letters of C. S. Lewis*. Grand Rapids, MI: Zondervan.

Huxley, Sir Julian. 1955. *Evolution and Genetics*, Chapter 8: *What Is Science?*, pp. 256-289. New York: Simon & Schuster. (Edited by James R. Newman)

Irvine, William. 1959. The Influence of Darwin on Literature. *Proceedings of the American Philosophical Society* **103**(5):616-628, October 15..

Janus, Christopher G. and William Brashler, 1975. *The Search for Peking Man*. New York: Macmillan.

Joshi, S. T. 2003. *God's Defenders*. Amherst, NY: Prometheus Books.

Kilby, Clyde S. 1964. *The Christian World of C. S. Lewis*. Grand Rapids, MI: Wm. B. Eerdmans.

_____. 1985. "Into the Land of Imagination." *Christian History* **4**(3):16-18.

Kroeber, Alfred L. 1920. "Totem and Taboo: An Ethnologic Psychoanalysis." *American Anthropologist* **22**(1):48–55, January-March.

Kuehn, K. 2004. *Brains, Minds, and Unicorns: A Critical Review of Victor Reppert's C. S. Lewis' Dangerous Idea*.<http://www.ps.uci.edu/~kuehn/personal/reppert.htm>.

Larson, Edward J. 1997. *Summer for the Gods: The Scopes Trial and America's Continuing Debate over Science And Religion*. New York: Basic Books.

_____. 2012. "C. S. Lewis on Science as a Threat to Freedom." Chapter 3 in West, 2012, pp. 53-58.

_____, and Larry Witham. 1998. Leading scientists still reject God. *Nature* **394**:313, July 23.

Latta, Corey. 2016. *C. S. Lewis and the Art of Writing: What the Essayist, Poet, Novelist, Literary Critic, Apologist, Memoirist, Theologian Teaches Us about the Life and Craft of Writing.* Eugene, OH: Wipf and Stock Publishers.

Lazo, Andrew and Mary Anne Phemister, (editors). 2009. *Mere Christians: Inspiring Stories of Encounters with C. S. Lewis.* Grand Rapids, MI: Baker Books.

Levine, Joseph S. and Kenneth R. Miller. 1994. *Biology: Discovering Life.* Second edition, Lexington, MA: D.C. Heath. p. 658.

Levi-Strauss, Claude. 1996. The Structural Study of Myth. In *The Continental Philosophy Reader*, pp. 305-327. Edited by Richard Kearney and Mara Rainwater. New York: Routledge.

Lewis, C(live) S(taples) 1942. *A Preface to Paradise Lost.* Oxford: Oxford University Press.

_____. 1943. *The Screwtape Letters.* New York: Macmillan.

_____. 1944. *The Pilgrims Regress.* New York: Sheed and Ward.

_____.1949. *The Weight of Glory and Other Addresses.* New York : Macmillan Co

_____. 1954. "The Great Divide." *Christian History* **4**(3):32.

_____. 1958. *Reflections on the Psalms.* New York: Harcourt Brace Jovanovich.

_____. 1958a. "Willing Slaves of the Welfare State." *The Observer*, July 20. http://liberty-tree.ca/research/willing_slaves_of_the_welfare_state

_____. 1960. *The World's Last Night and Other Essays.* New York: Harcourt Brace Jovanovich.

_____. 1960a. *Mere Christianity.* New York: Collier Publishing.

_____. 1962. *They Asked for a Paper: Papers and Addresses.* London: Geoffrey Bles. This collection includes his presentation to Cambridge titled *De Descriptione Temporum*

_____. 1964. *The Discarded Image: An Introduction to Medieval and Renaissance Literature*. Cambridge: Cambridge University Press.

_____. 1964a. *Letters to Malcolm: Chiefly on Prayer*. New York: Harcourt, Brace & World.

_____. 1965. *The Abolition of Man: Or, Reflections on Education*. New York: Macmillan.

_____. 1966. *Letters of C. S. Lewis*. Grand Rapids, MI: Eerdmans.

_____. 1967. *Christian Reflections*. Grand Rapids, MI: Eerdmans. (Edited by Walter Hooper.)

_____ 1967a. *Studies in Words. Second edition*. New York: Cambridge University Press.

_____. 1969. "*De Descriptione Temporum*" in *Selected Literary Essays*. Cambridge: Cambridge University Press.

_____. 1970. *God in the Dock: Essays on Theology and Ethics*. Grand Rapids, MI:
Eerdmans. (Edited by Walter Hooper.)

_____. 1974. *That Hideous Strength: A Modern Fairy-Tale for Grown-Ups.*. New York: Scribner, p. 39.

_____. 1977. *The Joyful Christian*. New York: Macmillan.

_____. 1980. *Mere Christianity*. New York: Simon & Schuster.

_____. 1980a. *The Weight of Glory and Other Addresses, Revised Edition*. New York: Macmillan.

_____. 1982. *C. S. Lewis: On Stories and Other Essays on Literature*. New York: Harcourt Brace Jovanovich.

_____. 1984. *The Business of Heaven: Daily Readings from C. S. Lewis*. London: Fount Paperbacks. (Edited by Walter Hooper.)

_____. 1986. *Present Concerns*. New York: Harcourt Brace Jovanovich. (Edited by Walter Hooper.)

_____. 1991. *All My Road Before Me: The Diary of C. S. Lewis, 1922-*

1927. San Diego: Harcourt Brace Jovanovich. (Edited by Walter Hooper.)

_____. 1992. *The Pilgrim's Regress: An Allegorical Apology for Christianity, Reason, and Romanticism*. Grand Rapids, MI: Eerdmans.

_____. 1993. *Letters of C. S. Lewis. Revised and Enlarged Edition*. New York: Harcourt.

_____. 1996. *The Collected Works of C. S. Lewis: The Pilgrim's Regress, Christian Reflections, God in the Dock*. New York: Inspirational Press.

_____. 1996a. *A Grief Observed*. New York: HarperCollins.

_____. 1996b. *Miracles*. New York: Simon & Schuster.

_____. 1996c. *Perelandra*. New York: Scribner. First Edition 1944.

_____. 1996d. *The Problem of Pain*. New York: HarperCollins.

_____. 1996e *That Hideous Strength: A Modern Fairy-Tale for Grown-Ups*. New York: Simon & Schuster. Originally published in 1945.

_____. 1996f. *The Weight of Glory*. New York: Simon & Schuster.

_____. 2001a. *Miracles: A Preliminary Study*. San Francisco: HarperSanFrancisco (now HarperOne).

_____. 2002. *Surprised By Joy: The Shape of My Early Life*. New York: Barnes & Noble Modern Classics Books.

_____. 2004. *The Chronicles of Narnia*. New York: HarperCollins.

_____. 2004a. *The Collected Letters of C. S. Lewis, Volume 1: Family Letters, 1905- 1931*.San Francisco: HarperSanFrancisco [now HarperOne].

_____. 2004b. *The Collected Letters of C. S. Lewis, Volume 2: Books, Broadcasts, and War, 1931-1949*.San Francisco: HarperSanFrancisco.

_____. 2007. *The Collected Letters of C. S. Lewis, Volume 3: Narnia. Cambridge, and Joy, 1950-1963*.San Francisco: HarperSanFrancisco. [Includes letter to Janet Wise from C. S. Lewis, dated October 5, 1955.]

_____. 2008. *Yours, Jack: Spiritual Direction from C. S. Lewis*. New York: HarperOne.

_____. 2012. *The World's Last Night and Other Essays*. New York: Mariner Books.

_____. 2013. *Studies in Medieval and Renaissance Literature*. Cambridge: Cambridge University Press.

Lewis, Gordon R. and Bruce A. Demarest. 2010. *Integrative Theology*. Grand Rapids, MI: Zondervan.

Line, Peter and Jerry Bergman. 2020. *The Claims for Evolution of Mankind Carefully Demolished*. Sudbury, MA: Bartlett Publishing.

Loomis, Steven R. and Jacob P. Rodriguez. 2009. *C. S. Lewis: A Philosophy of Education*. New York: Palgrave Macmillan.

Markos, Louis. 2003. *Lewis Agonistes: How C. S. Lewis Can Train Us to Wrestle with the Modern and Postmodern World*. Nashville, TN: Broadman & Holman Publishers.

_____. 2012. E-Mail to Jerry Bergman, dated September 12.

_____ and David Diener. 2015. *C. S. Lewis: An Apologist For Education*. Camp Hill, PA: Classical Academic Press.

Martindale, W., and J. Root. (editors). 1990. *The Quotable Lewis*. Wheaton, IL: Tyndale House Publishers, Inc.

Mathison, Jane. 1979. "Darwin & the Death of Natural Theology." *The Teilhard Review* **14**(1):23-35.

McCulloch, Winifred. 1996. *Teilhard de Chardin and the Piltdown Hoax*. Teilhard Studies Number 33. Lewisburg, PA: Bucknell University. http://teilharddechardin.org/old/studies/33-Teilhard_and_the_Piltdown_Hoax.pdf

McDermott, Robert A. 1995. "Rudolf Steiner and Anthroposophy", in Faivre and Needleman, *Modern Esoteric Spirituality*, New

York: Crossroad Publishing.

McGrath, Alister. 2013. *C. S. Lewis: A Life*. Carol Stream, IL: Tyndale.

Miller, Ryder W. (editor). 2003. *From Narnia to A Space Odyssey*. New York: iBooks.

Moore, James. 1979. *The Post Darwinian Controversies*. New York: Cambridge University Press.

Murphy, Brian. 1983. *C .S. Lewis (Starmont Reader's Guide, 14)*. Mercer Island, WA: Starmont House.

Myers, Doris. 1994. *C. S. Lewis in Context*. Kent, OH: Kent State University Press.

Myers, Ellen. 1985. "Creationist Gleanings from C. S. Lewis." Reprinted from *Creation Social Science and Humanities Society* (Quarterly Journal) **8**(1):10-13, Fall.

Nicholi, Armand M. 2002. *The Question of God. C. S. Lewis and Sigmund Freud Debate God, Love, Sex, and the Meaning of Life*. New York: The Free Press.

Numbers, Ronald L. 2006. *The Creationists: From Scientific Creationism to Intelligent Design*. Cambridge: MA: Harvard University Press.

O'Leary, Denyse. 2004. *By Design or by Chance?* Ontario, Canada: Castle Quay Books.

Pelser, Adam C. 2017. "The Abolition of Man Today." *CHRISTIAN RESEARCH JOURNAL* **40**(2), April 24.

Pearce, Joseph. 2003. *C. S. Lewis and the Catholic Church*. San Francisco, CAS : Ignatius Press.

_____. 2013. *C. S. Lewis and the Catholic Church*. Charlotte, NC: Saint Benedict Press

Peterson, Michael L. 2010. "C. S. Lewis on Evolution and Intelligent Design." *Perspectives on Science and Christian Faith* **62**(4):253-266, December.

_____. 2020. *C. S. Lewis and the Christian Worldview*. New York:

Oxford University Press.

Poe, Harry Lee and Rebecca Whitten Poe, (editors). 2006. *C. S. Lewis Remembered: Collected Reflections of Students, Friends & Colleagues.* Grand Rapids, MI: Zondervan.

Poythress, Vern S. 2006. *Redeeming Science: A God-Centered Approach.* Wheaton, IL: Crossway Books.

Pranavananda, Swami. 1957. "The Abominable Snowman." *Journal of the Bombay Natural History Society* **54**(1-2):179-181.

Purtill, Richard L. 1981. *C. S. Lewis' Case for the Christian Faith*, First edition. San Francisco, CA: Harper & Row; 2004 edition, San Francisco, CA: Ignatius Press.

Rainer, Tom S. 2001. *Surprising Insights from the Unchurched and Proven Ways to Reach Them.* Grand Rapids, MI: Zondervan.

Raymond, E. T. 1920. *A Life of Arthur James Balfour..* Boston: Little Brown.

Reddington, Kenneth G. 2015. Following the Truth, Wherever It Leads: An Investigation of What Is Reality (and How It Affects Our Lives). Eugene, OR: Wipf and Stock.

Reid, Daniel G. 1990. *Dictionary of Christianity in America.* Downers Grove, IL: InterVarsity Press.

Reppert, Victor. 2003. *C. S. Lewis' Dangerous Idea.* Downers Grove, IL: InterVarsity Press.

Roberts, Adam. 1998. *Silk and Potatoes: Contemporary Arthurian Fantasy.* Atlanta, GA: Rodopi.

Ruse, Michael. 1979. *The Darwinian Revolution: Science Red in Tooth and Claw.* Chicago: University of Chicago Press.

_____. 2001. *The Evolution Wars: A Guide to the Debates.* New Brunswick, NJ: Rutgers University Press.

_____. 2017. *Darwin as Religion.* New York: Oxford University Press.

_____. 2019. *A Meaning to Life.* New York: Oxford University Press.

Sagan, Carl. 1980. *Cosmos*. New York: Random House.

Santamaria, Abigail. 2015. *Joy: Poet, Seeker, and the Women Who Captivated C. S. Lewis*. Boston: Houghton Mifflin Harcourt.

Sarfati, Jonathan. 2015. *The Genesis Account: A Theological, Historical and Scientific Commentary on Genesis 1-11*. Powder Springs, GA: Creation Book Publishers.

Sayer, George. 1994. *Jack: A Life of C. S. Lewis*. Wheaton, IL: Crossway Books, pp. 176-184.

Schmerl, Rudolf B. 1960. "Reason's Dream: Anti-Totalitarian Themes and Techniques of Fantasy."

Schultz, Jeffrey and John G. West. 1998. *Encyclopedia of Religion in American* Politics. Westport, CT: Greenwood Press.

Schumacher, Leo S. 1968. *The Truth About Teilhard*. New York: Twin Circle Publishing Company.

Schwartz, Sanford. 2009. *C. S. Lewis on the Final Frontier: Science and the Supernatural in the Space Trilogy*. New York: Oxford University Press.

Shermer, Michael. 2000. *How We Believe*. New York: Freeman.

Smilde, Arend. 2014. "C. S. Lewis, St. Jerome, and the Biblical Creation Story: The Background of a Recurring Misattribution." *Journal of Inklings Studies* **4**(2):115-124, October.

Smith, F. LaGard. 2018. *Darwin's Secret Sex Problem: Exposing Evolution's Fatal Flaw—the Origin of Sex*. Bloomington, IN: WestBow Press.

Teilhard de Chardin, Pierre. 1979. *The Heart of Matter*. New York: Harcourt Brace Jovanovich, p. 25. (Translated from the French by René Hague.)

Tolson, J. 2005. "God's Storyteller: The Curious Life and Prodigious Influence of C. S. Lewis, the Man Behind *The Chronicles of Narnia*." *U.S. News & World Report* **139**(22): 46-52, December 12.

Tomkins, Jeffrey P. and Jerry Bergman. 2012. "Is the human genome nearly identical to chimpanzee?—a reassessment of the literature." *Journal of Creation* **25**(4):54–60.

Turner, J. Scott. 2007. *The Tinker's Accomplice: How Design Emerges From Life Itself.* Cambridge, MA: Harvard University Press.

Walsh, Chad. 1949. *C. S. Lewis: Apostle to the Skeptics*. New York: Macmillan.

Ward, Michael. 2013. "Science and Religion in the Writings of C. S. Lewis." *Science & Christian Belief* **25**(1):3-16, April. A lecture hosted by The Faraday Institute, St. Edmund's College, Cambridge on Tuesday, 29 May 2012.

Watson, D. M. S. 1929. "Adaptation." *Nature* **124**(3119):231-234, August 10.

Weikart, Richard. 2012. "C. S. Lewis and Science." *Credo Magazine*, October 24.

http://www.credomag.com/2012/10/24/c-s-lewis-and-science.

_____. 2019. Whatever Happened to Human Rights?: Morality after C. S. Lewis' Abolition of Man. *Christian Research Journal*, December 17. https://www.equip.org/article/whatever-happened-to-human-rights-morality-and-c-s-lewiss-abolition-of-man/

West, John G. (editor). 2012. *The Magician's Twin: C. S. Lewis on Science, Scientism, and Society*. Seattle, WA: Discovery Institute Press.

Whilk, Nat (pseudonym for C. S. Lewis). 1957. "Evolution Hymn." *The Cambridge Review* **78**:227, November 30.

Whiston, William (Translator). 1987. *The Works of Flavius Josephus*. Peabody, MA: Hendrickson Publishers.

White, William Luther. 1969. *The Image of Man in C. S. Lewis*. Nashville, TN: Abingdon Press.

Wielenberg, Erik J. 2008. *God and the Reach of Reason: C. S. Lewis, David Hume, and Bertrand Russell*. New York: Cambridge University Press.

Wile, Jay. 2011. "Thoughts from a Scientist who is a Christian (Not a Christian Scientist)." *Everyone Wants a Piece of C. S. Lewis.* Friday, July 22.

Williams, Charles, and C. S. Lewis. 1974. *Taliessin through Logres [and] The Region of the Summer Stars, by Charles Williams. And Arthurian Torso.* Grand Rapids, MI: Eerdmans. (Introduction by Mary McDermott Shideler.)

Williams, David. 2012. "Surprised by Jack: C. S. Lewis on *Mere Christianity*, the Bible, and Evolutionary Science." *BioLogos*, December 10. https://biologos.org/people/david-williams

Williams, Donald T. 2006. *Mere Humanity: G.K. Chesterton, C. S. Lewis, and J. R. R. Tolkien on the Human Condition.* Nashville, TN: B&H Books.

Wilson, A. N. 1990. *C. S. Lewis: A Biography.* New York: Norton.

Yancey, Philip. 2010. *What Good is God?* New York: Faith Words.

Appendix 1

A View from Another C. S. Lewis Fan

Ellen Myers. 1985. "Creationist Gleanings from C. S. Lewis." Reprinted from *Creation Social Science and Humanities Society* (Quarterly Journal) **8**(1):10-13, Fall.

I RELUCTANTLY BELIEVED for a number of years that C. S. Lewis, apostle and teacher to modern skeptics whom I greatly cherish, had made a kind of uneasy truce with modern evolutionism. My opinion was based on passages in **Mere Christianity**[1] and **The Problem of Pain**[2] where he drew upon evolutionist theories to illustrate his own explanations of Christian beliefs. It is certainly evident that the scope of Lewis' Christian writings is far wider than the specific area of origins by God's creation. Christians and unbelievers inquiring about the Christian faith find in Lewis the sorely needed definition and defense of "mere Christianity" of what has been part of Christianity at all times and in all its branches, and which sets Christianity proper apart from unbelief. C. S. Lewis deserves a special word of gratitude for his intransigence towards "modernism" posing as Christianity. The following statement made in answer to Sherwood E. Wirt of the Billy Graham Evangelistic Association about six months before Lewis' death is typical:

- **Mr. Wirt**: What is your opinion of the kind of writing being done within the Christian church today?

- **Lewis**: A great deal of what is being published by writers in the religious tradition is a scandal and is actually turning people away from the church. The liberal writers who are continually accommodating and whittling down the truth of the Gospel are responsible. I cannot understand how a man can appear in print claiming to disbelieve everything that he presupposes when he puts on the surplice [ecclesiastical clothing]. I feel it is a form of prostitution.[3]

It always seemed to me that one as committed to the truth of the Gospel as C. S. Lewis could not have been neutral on the issue of creation ex nihilo versus evolutionism, but would have welcomed the vindication of the doctrine of creation we are observing today, as did Malcolm Muggeridge in no uncertain terms.[4] That vindication, however, was for the most part still in the future during Lewis' life; we must remember that when he died in 1963, *The Genesis Flood*, the creationist classic which first put modern creation science before the general and academic public, had been in print for barely two years. Nevertheless there are certain Lewis writings indicating that he was indeed on the creationist side. I shall make reference to some of these writings, advising the reader that my sampling is not intended to be exhaustive.

Lewis debunks the modern arrogant unbelief based upon "evolution and comparative religion... and all the guess-work which masquerades as 'science'" as early as in his largely autobiographical *The Pilgrim's Regress*,[5] the first book published after his conversion. He himself made it clear that he had evolutionism in mind when he put the following words in the mouth of his foolish old "Mr. Enlightenment":

Hypothesis... establishes itself by a cumulative process, or, to use proper language, if you make the same guess often enough it ceases to be a guess and becomes a Scientific Fact.[6]

How appropriate these sarcastic words are now, fifty years later, about evolutionist 'dogma'. (I use the word "dogma" deliberately in the sense Lewis once said that common everyday people use it "in a bad sense to mean 'unproved assertion delivered in an arrogant manner.'"[7] Common everyday people have a good deal of common sense!)

Lewis' essay "Two Lectures," originally published in February 1945, can only be called a bit of creationist apologetics. It contrasts an urbane lecture about "Evolution, development, the slow struggle upwards and onwards from crude and inchoate beginnings towards ever-increasing perfection and elaboration that appears to be the very formula of the whole Universe"[8] with a Dream Lecturer pointing out that "(t)he acorn comes from a full-grown oak. The Rocket comes, not from a still crude engine, but from something much more perfect than itself and much more complex, the mind of a man, and a man of genius. The march of all things is from higher to lower..."[9] Lewis reaches the conclusion that

> 'Developmentalism' is made to look plausible by a kind of trick… And since the egg-bird-egg sequence leads us to no plausible beginnings, is it not reasonable to look for the real origin somewhere outside sequence altogether? You have to go outside the sequence of engines, into the world of men, to find the real originator of the Rocket. Is it not equally reasonable to look outside Nature for the real Originator of the natural order?[10]

Perhaps Lewis' most explicit anti-evolutionist writing is found in his essay, "The Funeral of a Great Myth." It was published for the first time four years after Lewis' death (not surprising in view of the then still prevailing climate of evolutionist dogmatism) in the collection *Christian Rellections*.[11] While the bulk of the article deals with the mythological implications of evolutionism in a manner devastating to the makers and believers of such myths, Lewis also questions the scientific validity of evolutionism. He makes it very clear that even to the biologist, evolution is a hypothesis; and he speculates about the grounds for accepting the hypothesis as largely metaphysical and the fulfillment of an "imaginative need."[12] Lewis correctly predicts that the evolutionist hypothesis, even strictly on scientific evidences, "may be shown, by later biologists, to be a less satisfactory hypothesis than was hoped fifty years ago."[13] He also States that "probably every age gets, within certain limits, the science it desires."[14] In this Statement he anticipated the later admissions of philosophers of science, for instance Imre Lakatos, that there is no pure objectivity in scientific research.[15]

Throughout Lewis' writings runs the idea that in the course of their continued existence all things or persons only become more and more themselves. "Even on the biological level," he writes in ***The Great Divorce***, "life is not like a pool but like a tree. It does not move towards unity but away from it and the creatures grow further apart as they increase in perfection."[16] While the notion of a "tree" of biological development is all too familiar to us through the popular evolutionist "tree" model (now amended to a "bush" by the "punctuated equilibrium" model proposed by S.J. Gould and Niles Eldredge) Lewis' emphasis was upon differentiation, and certainly not upon the notion of one species, kind or individual changing into another. Such trans-species change from boy into dragon or from Talking Beast to common wild beast as in Lewis' beautiful ***Narnia*** stories is always supernatural and ominous, revealing latent evil in his characters. Such changes are also always downward The same idea of greater and stronger individual separateness and perfected unique identity is also found in the life of the heroine Orual in ***Till We Have Faces***, or in a conversation between good Cecil and Margaret Dimble in ***That Hideous Strength***. Such increasing individual identity which becomes ultimately fixed as a result of God's creative design and the individual's choice runs counter to random naturalistic emergent evolutionism in every fundamental and operational respect.

No overview of this matter would be complete without reference to Lewis' portraits of lost and evil men in his science fiction trilogy, **Out of the Silent Planet**, **Perelandra**, and **That Hideous Strength**. From the cold contempt and unscrupulous exploitation of feebleminded Harry by Devine and Weston in Planet, to Filostrato's experiment with a guillotined man's head and Wither's trancelike senility in a demon-made void in Strength there is a wealth of prophetic realism about the end result of the emergent evolutionist world view for its practitioners-victims. Most horrible of them all is Weston turned "UnMan" in **Perelandra**. He is "a convinced believer in emergent evolution. All is one."[17] Weston believes that "pure spirit.... is the goal towards which the whole cosmic process is moving.... Call it a Force. A great, inscrutable Force, pouring up into us from the dark bases of being... your Devil and your God are both pictures of the same Force."[18] Anyone acquainted with the emergent evolutionist philosophy of the maverick Catholic priest Teilhard de Chardin cannot help wondering whether Lewis modeled his Weston upon Teilhard.

We may conclude from all these excerpts that Lewis was unalterably opposed to emergent evolutionism as a philosophy or myth; that he maintained a good deal of dry skepticism about the biological hypothesis of evolutionism as not necessarily scientifically true; and that he would have gladly welcomed the rise of creation science in our own generation.

FOOTNOTES

(Titles below are by C. S. Lewis unless otherwise indicated.)

1. cf. Mere Christianity (New York: Macmillan Publishing Co., Twenty-Ninth Paperback Printing 1979), 1 84ff.
2. cf. The Problem of Pain (New York: Macmillan Publishing Co., Thirteenth Paperback Printing 1972), 77ff.
3. God in the Dock (Grand Rapids, Ml: William B. Eerdmans Publishing Co., 1970), 260.

4. "Muggeridge on Evolution," in Creation Social Science and Humanities Quarterly. Vol. III, No. 4 (Summer 1981), 17.
5. The Pilgrim's Regress (Grand Rapids, Ml: Wm. B. Eerdmans Publishing Co., 1933, Reprinted November 1971), 36, 37 (headlines).
6. Ibid, 37.
7. God in the Dock, 97.
8. Ibid, 208.
9. Ibid, 209.
10. Ibid. 210, 211.
11. Christian Reflections (Grand Rapids, MI: William B. Eerdmans Publishing Co. 1967)82-93. Also see Preface by Walter Hooper, p. xiii, for publication date.
12. Ibid, 85.
13. Ibid. 83.
14. Ibid, 85.
15. cs. Imre Lakatos and Alan Musgrave, editors, Criticism and the Growth of Knowledge (Cambridge University Press, 1 970), pp. 74, 92, 99.
16. The Great Divorce (New York: The Macmillan Company, 1946, Sixteenth Printing 1970), vi.
17. Perelandra (New York: The Macmillan Company, Tenth Paperback Printing 1970) 90.
18. Ibid, 92, 93.

About the Author

Jerry Bergman, Ph.D.

Dr. Bergman taught biology, chemistry, and anatomy at Northwest State College in Archbold, Ohio and is an Adjunct Associate Professor at the University of Toledo Medical College. He has 9 earned degrees, including 5 graduate degrees and a Ph.D. from Wayne State University in Detroit, Michigan. His over 1,800 publications are in both peer-reviewed scholarly and popular science journals. Dr. Bergman's work has been translated into 13 languages including French, German, Italian, Spanish, Danish, Arabic, Polish, and Swedish.

His books – as well as the books that include chapters he has authored – are in over 1,400 college and major public libraries in 26 countries. So far, over 80,000 copies of the 48 books and monographs that he has authored or co-authored are in print. He has spoken over 2,000 times to college, university, and church groups in America, Canada, Europe, Asia, and Africa. He has also been a guest on hundreds of radio and television shows.

About the Cántaro Institute
Inheriting, Informing, Inspiring

Cántaro Institute is a reformed evangelical organization committed to the advancement of the Christian worldview for the reformation and renewal of the church and culture.

We believe that as the Christian church returns to the fount of Scripture as her ultimate authority for all knowing and living, and wisely applies God's truth to every aspect of life, faithful in spirit to the reformers, her missiological activity will result in not only the renewal of the human person but also the reformation of culture, an inevitable result when the true scope and nature of the gospel is made known and applied.

www.ingramcontent.com/pod-product-compliance
Lightning Source LLC
Chambersburg PA
CBHW030248010526
44107CB00031B/1367/J